Engineering Drawing, Communication, and Design

PETER COOLEY

C.Eng., M.I.Mech.E.
Department of Mechanical Engineering
University of Aston in Birmingham

Pitman Publishing

First published 1972

SIR ISAAC PITMAN AND SONS LTD.
Pitman House, Parker Street, Kingsway, London, WC2B 5PB
P.O.Box 6038, Portal Street, Nairobi, Kenya

SIR ISAAC PITMAN (AUST.) PTY. LTD.
Pitman House, Bouverie Street, Carlton, Victoria 3053

PITMAN PUBLISHING COMPANY S.A. LTD.
P.O.Box 9898, Johannesburg, S.Africa

PITMAN PUBLISHING CORPORATION
6 East 43rd Street, New York, N.Y.10017, U.S.A.

SIR ISAAC PITMAN (CANADA) LTD.
Pitman House, 381-383 Church Street, Toronto, 3, Canada

THE COPP CLARK PUBLISHING COMPANY
517 Wellington Street, Toronto, 2B, Canada

A/604.2'4

Cased Edition   ISBN 0 273 43940 5

Paperback Edition   ISBN 0 273 31798 9

G2-(T199/523:76)

B/5/2 M.   £2.70

Engineering Drawing, Communication, and Design

Preface

There is an old practical joke in which a group of people who are "in the know" ask a newcomer to describe a spiral staircase. Very often the newcomer does as expected and traces out the locus of the stairs with a finger whilst saying "It's a staircase that goes round like this". The rest of the group are prepared for this and can laugh at his predictable behaviour. The trick works because some objects are very difficult to describe in words and most people resort to pictures or models when they have to deal with them in detail.

The engineering designer has to cope with this problem throughout most of his professional life. In order to communicate ideas and instructions he uses sketches and drawings as well as written and spoken words. There are many books on engineering drawing and on the various aspects of engineering design but little has been written which shows the relationship between the two activities. Surely this is a very strange state of affairs? Engineering drawings and sketches are the engineering designer's special method of communicating his ideas to other people but why should the study of this communication be isolated from the process which uses it? In this book I have attempted to treat the basic topic of drawing and sketching and then to show how they may be used in the synthesis of a new product.

The earlier chapters cover most of the essential engineering drawing background required by most students of engineering. They can be used either as a text or as a self-instruction manual. To facilitate the latter, solutions are provided for many of the exercises. Whenever possible the exercises are design-orientated: students choose the most suitable views of a given object rather than being told exactly what to draw. The units used are SI throughout.

The last eight chapters illustrate the design process. It is extremely difficult to present a realistic project in a book; only in an industrial environment are there real constraints on the designer which could be easily appreciated. However, I believe that the unusual and, frankly, experimental scripted case-history approach will prove stimulating to most readers. My own experience with this technique has been most encouraging and I hope that other students of engineering design will find it interesting and informative.

I would like to thank Mr. A. Dunsmore for
several helpful suggestions, my wife for her help
and encouragement, and my children for not making
too much noise.

Peter Cooley

Streetly, Staffs.

September 1971.

A/604.2'4

Contents

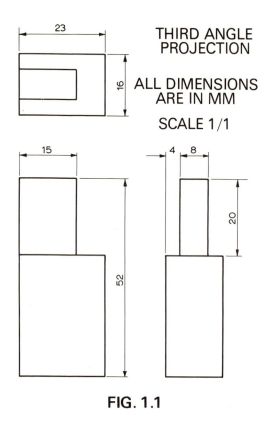

THIRD ANGLE PROJECTION

ALL DIMENSIONS ARE IN MM

SCALE 1/1

FIG. 1.1

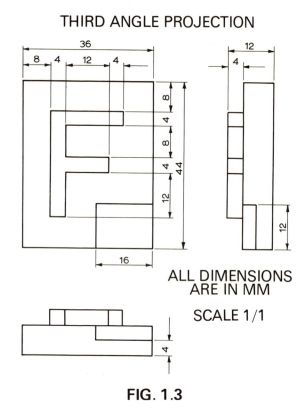

THIRD ANGLE PROJECTION

ALL DIMENSIONS ARE IN MM

SCALE 1/1

FIG. 1.3

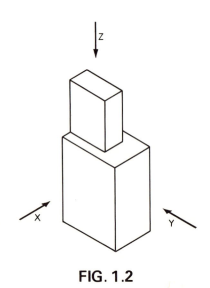

FIG. 1.2

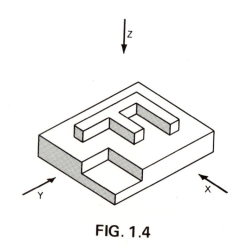

FIG. 1.4

CHAPTER 1

Orthographic Projection

Figure 1.1 is an engineering drawing of a simple, solid object but it contains many of the features of all working drawings. The first thing to notice is that three separate views of the object have been drawn. Study the views carefully and try to picture the solid object. Make a sketch of the object and then insert arrows to indicate the directions in which the views of Fig.1.1 could be seen.

You will have made some assumptions about the arrangement of the views in Fig.1.1; can you justify these assumptions by relating Fig.1.1 to your sketch? When you have checked on this point, compare your sketch with *Figure 1.2*. Your sketch may be drawn as seen from a viewing point different from that in Fig.1.2, but the important thing is to have the same concept of the object as the figure illustrates.

In an engineering drawing, the exact shapes of the surfaces of the object are constructed with reference to two or more perpendicular planes. In the above example, there are no curved surfaces to complicate matters and every surface is rectangular. The object is referred to imaginary planes that are parallel to the surfaces of the object and, therefore, every surface appears in its true shape. The distances need not be full size, although it is often convenient to make them so.

Fig.1.9 is to a scale of 1/2, i.e. 1 mm on the paper represents 2 mm on the object.

Figs. 1.1 and 1.2 illustrate how a solid object is represented in Orthographic Projection. By accurately drawing out the surfaces that appear when one looks in the direction of arrow X, the lower left-hand view (of Fig.1.1) would be produced. Similarly, arrow Y indicates the source of the lower right-hand view and arrow Z indicates the source of the upper view.

It must be emphasized that the views contain the true shapes of the surfaces. It is neither necessary nor desirable to employ the rules of perspective in order to draw what a person (having two eyes, about 60 mm apart) would actually see from a particular viewing point.

Figure 1.3 shows three views of a more complicated object than the one above. Try to visualize the solid object and then make a sketch of it. Insert arrows to indicate the directions in which the views of Fig.1.3 could be seen. If you have difficulty, try referring back to Figs. 1.1 and 1.2. Although the views in Figs. 1.1 and 1.3 are arranged differently, the relationship between the views is the same; it is known as THIRD ANGLE PROJECTION and it will be explained below.

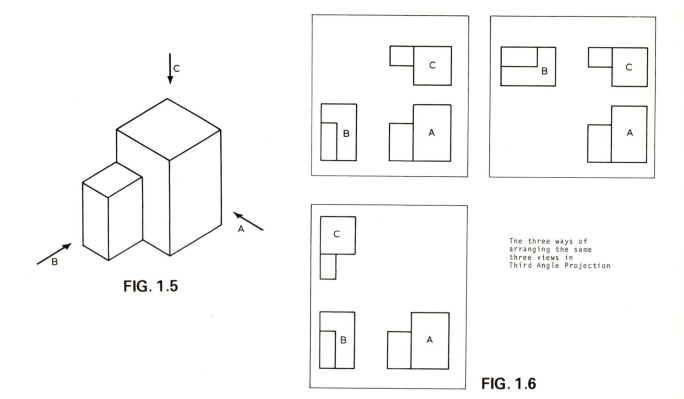

FIG. 1.5

The three ways of
arranging the same
three views in
Third Angle Projection

FIG. 1.6

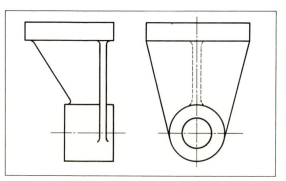

THIRD ANGLE PROJECTION

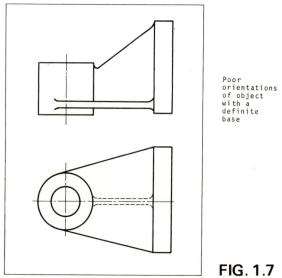

Poor
orientations
of object
with a
definite
base

FIG. 1.7

When you have finished your sketch, compare it with *Figure 1.4*. Your sketch may be drawn as seen from a viewing point different from that in Fig.1.4, but it should be, essentially, the same object. By accurately drawing out the surfaces that appear when one looks in the direction of arrow X, the upper right-hand view (of Fig.1.3) would be produced. Notice that the long (44 mm) edge has been placed vertically. Arrow Y indicates the source of the lower view and arrow Z indicates the source of the upper left-hand view.

Why have the views been positioned as they have in Figs. 1.1 and 1.3? In other words, can you work out the rule for positioning the views called THIRD ANGLE PROJECTION? Write it down and then compare your answer with the explanation given below.

The rule is very simple. Pick any of the views in Figs. 1.1 and 1.3 (this view will be referred to as the "selected" view). What would be seen of the solid object, from the <u>left</u> of the selected view, is drawn, horizontally, to the <u>left</u>. What would be seen from the <u>right</u> of the selected view is drawn, horizontally, to the <u>right</u>. What would be seen from <u>above</u> the selected view is drawn, vertically, <u>above</u>. What would be seen from <u>below</u> the selected view is drawn, vertically, <u>below</u>.

Corresponding surfaces line up along vertical and horizontal lines: this is of considerable assistance to the person making the drawing and to the person reading the drawing.

Examine all the views of Figs. 1.1 and 1.3 and you will find that the rule holds good. Notice that there is one view which links the other two together, because it has views projected both horizontally and vertically from it. This view is dubbed the *Link View* and it will be shown that, in any drawing with three views from mutually perpendicular directions, any of the views may act as the link view in third angle projection.

Figure 1.5 is a pictorial view of a simple block. Views are required as seen in the directions of arrows A, B and C; they are to be arranged in third angle projection. Any one of the views can be the link view. Use a sheet of graph paper and estimate the proportions of the object, then sketch out the three views with the view in the direction of arrow A as the link view.

Now sketch out the same three views, but make the view in the direction of arrow B the link view. It is still possible to follow the rule for third angle projection.

Thirdly, sketch out the same three views but with the view in the direction of arrow C acting as the link view. When you are satisfied that the sets of views are all in accordance with the rule for third angle projection, compare your sketches with those in *Figure 1.6*.

You may have a combination that looks different from those shown in the figure. Try turning your paper through 90° or 180° and you should see that it is the same as one of those in Fig.1.6.

It is not always easy to decide upon which arrangement to use for an Engineering Drawing; much depends on the object in the drawing and on the shape of the paper on which the views are to be drawn. Some objects have surfaces and edges that would look very strange if they were not orientated in the normal way. For example, a house would look most peculiar if its vertical walls were shown horizontally in a drawing. A very long object looks strange if its length is not along the horizontal, and there may be some difficulty in fitting it on the paper. Other objects, like the one shown in *Figure 1.7*, have an obvious base which should be drawn in the normal position, and not with the object appearing to stand on its head or its side. However, an object, like the one in Fig.1.5, could be drawn out in any of the possible arrangements consistent with the type of projection used; a draughtsman would use the one that he found most convenient.

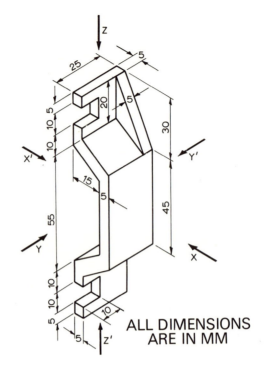

ALL DIMENSIONS
ARE IN MM

FIG. 1.8

THIRD ANGLE PROJECTION

SCALE 1/2

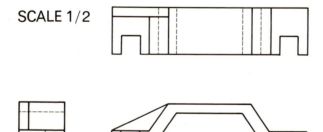

FIG. 1.9

EXERCISE 1.1

Figure 1.8 is a pictorial view of a bracket. Views, as seen in the directions of arrows X, Y and Z, are to be used for an engineering drawing. Why are these views better than those in the directions of arrows X', Y' and Z'?

Draw the views out, FULL SIZE (a scale of 1/1), in third angle projection. Choose a link view so that the object does not appear with an unnatural orientation. Make the drawing on graph paper; do not attempt to use instruments at this stage. Do *not* put any dimensions on your drawing; dimensioning is a subject in its own right and is dealt with in Chapter 5.

When you have finished, compare your drawing with *Figure 1.9*.

First and third angle projection

So far, the examples in this Chapter have dealt exclusively with THIRD ANGLE PROJECTION. This is the system of orthographic projection used throughout the United States of America, in several Commonwealth countries, and in many British industries. The other system used in Britain (and in most of Europe) is known as FIRST ANGLE PROJECTION. It is the exact opposite of third angle projection, i.e. what is seen from above a selected view is drawn vertically below that view; what is seen from the left of the selected view is drawn horizontally to the right of that view, etc.

The British Standards Institution issues its recommendations for Engineering Drawing Practice in British Standard 308 (B.S.308). Both First Angle Projection and Third Angle Projection are acceptable as British Standards. Both systems have their supporters but it cannot be proved that, for a trained person, one method has any advantages over the other. However, there is some evidence[1] that a person who has never studied an engineering drawing before will tend to assume that the views are arranged in third angle projection.

For this reason, third angle projection will be used for explaining each new principle but, whenever space permits, both types of projection will be introduced for the exercises.

Because there are two methods of projection in use in Britain, it is important that drawings are clearly labelled to indicate whether FIRST or THIRD ANGLE PROJECTION has been used.

Lines and dimensions

Fig.1.3 illustrates several of the other recommendations contained in B.S. 308. Notice that, for the lines used in the drawing, there are two distinct thicknesses. Thick lines are used for visible outlines and thin lines are used for (amongst other purposes) the dimension lines (those with arrowheads) and the projection lines (those touched by the arrowheads). Thick lines should be from two to three times the thickness of thin lines. All the lines in Fig. 1.3 are continuous but other types will be encountered later on.

The dimensions are expressed in millimetres and there is a note on the drawing to make this clear. Centimetres are not recommended units for engineering drawings; millimetres are satisfactory for most detail drawings, without using a larger or a smaller unit.

Pictorial and orthographic drawings

Engineering drawings are normally made in first or third angle orthographic projection, which means that the views are drawn as seen in two or more perpendicular directions. There are always two views at least and these have to be considered together in order to visualize the object. An engineering drawing must also provide information for the manufacture of the object: namely, as dimensions and notes. It is too risky to allow the dimensions to be scaled from the drawing. However, it would be possible to put the dimensions around a pictorial view of the object but such a method is rarely used.

[1] SPENCER, J., Experiments on Engineering Drawing Comprehension, *Ergonomics*, vol.8, pp.93-110, January 1965

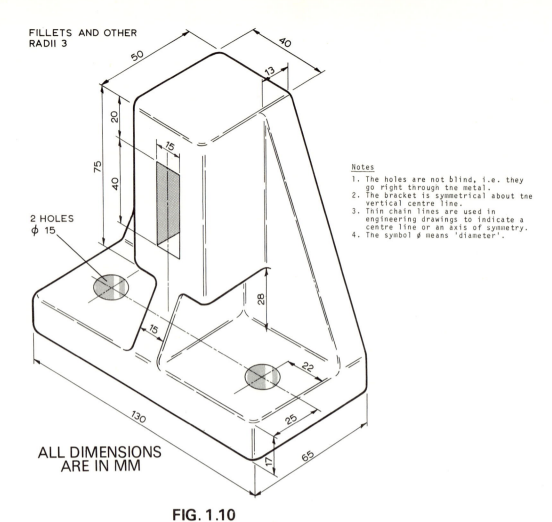

FILLETS AND OTHER RADII 3

2 HOLES ∅ 15

ALL DIMENSIONS ARE IN MM

Notes
1. The holes are not blind, i.e. they go right through the metal.
2. The bracket is symmetrical about the vertical centre line.
3. Thin chain lines are used in engineering drawings to indicate a centre line or an axis of symmetry.
4. The symbol ∅ means 'diameter'.

FIG. 1.10

THIRD ANGLE PROJECTION

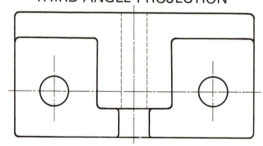

SCALE 1/2

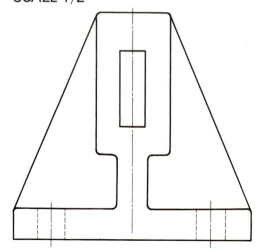

FIG. 1.11

Figure 1.10 illustrates some of the disadvantages of dimensioned pictorial views. The first point to notice is that, because the bracket is made by a casting process, it has very few sharp edges. The rounded corners, together with the circular holes and the general geometry of the object, make it quite awkward to draw. The shading has deliberately not been done very skillfully, but some shading is essential for indicating the true shape of the object.

Therefore, even a simple bracket like this one would be difficult and expensive to illustrate in this way. Once the object has been drawn, the dimensions have to be added. Notice that, although the bracket's geometry is quite straightforward, the dimensions take up a lot of space, are not always easy to follow, and confuse the pictorial quality of the drawing. A more complicated object would be extremely difficult to draw and to dimension and the cost of doing so could be prohibitive.

The advantages of orthographic projection are that drawings may be prepared quickly and cheaply by a person with no particular artistic skill and the views may be dimensioned with little difficulty and confusion. The disadvantage of orthographic projection is that one has to acquire some skill in visualizing the object in order to read the drawing. This is not a very serious disadvantage as most people can, with practice, develop the ability to understand orthographic projection.

EXERCISE 1.2

Which views will best illustrate the bracket shown in *Figure 1.10*? Are three views needed, or will two be sufficient? If you cannot decide on this point, look quickly at Fig.1.11, but use it only as a guide for which views to draw.

Decide on a suitable scale and plan the layout of the views *before* you begin to draw. Do not be caught with too little space left for one of the views. Use millimetre graph paper, a ruler and a coin to make a reasonable engineering drawing in third angle projection. When you have finished, compare your solution with Fig.1.11

Clarity in engineering drawings

Figure 1.11 shows two views, in third angle projection, of the bracket illustrated in Fig.1.10. Only two views have been drawn and these show the circular holes and the rectangular hole quite clearly. A third view would add nothing to the clarity of the drawing and, for this fairly simple object, would not help to distribute the dimensions more evenly.

Notice that every thick line on the drawing represents either the boundary between a metal surface and the surrounding air, or the intersection of two metal surfaces. At first sight there may appear to be some lines missing, but a comparison with Fig.1.10 will reveal that the line is missing because there is a radius that blends two surfaces smoothly. There are very few sharp edges and it is necessary to sketch all the circular arcs or to use some form of template.

As both views are symmetrical about the vertical centre line, it is convenient to work from this line and to transfer equal distances from one side to the other.

In addition to the requirements of Exercise 1.2, a new type of line has been used in Fig.1.11 to indicate the boundaries of the holes. It is convenient to show hidden details of this kind with <u>thin</u> short dashes and many engineering drawings include such lines.

The centre lines of the circular holes are shown in both views: another aid to clarity. Remember that centre lines, like dimension and hidden detail lines, should be thin lines; visible, thick lines are, at least, twice as thick as the thin lines. Drawings are much more difficult to read if this distinction is not maintained.

There are two labels in Fig.1.11 which should appear on every engineering drawing: the type of projection and the scale.

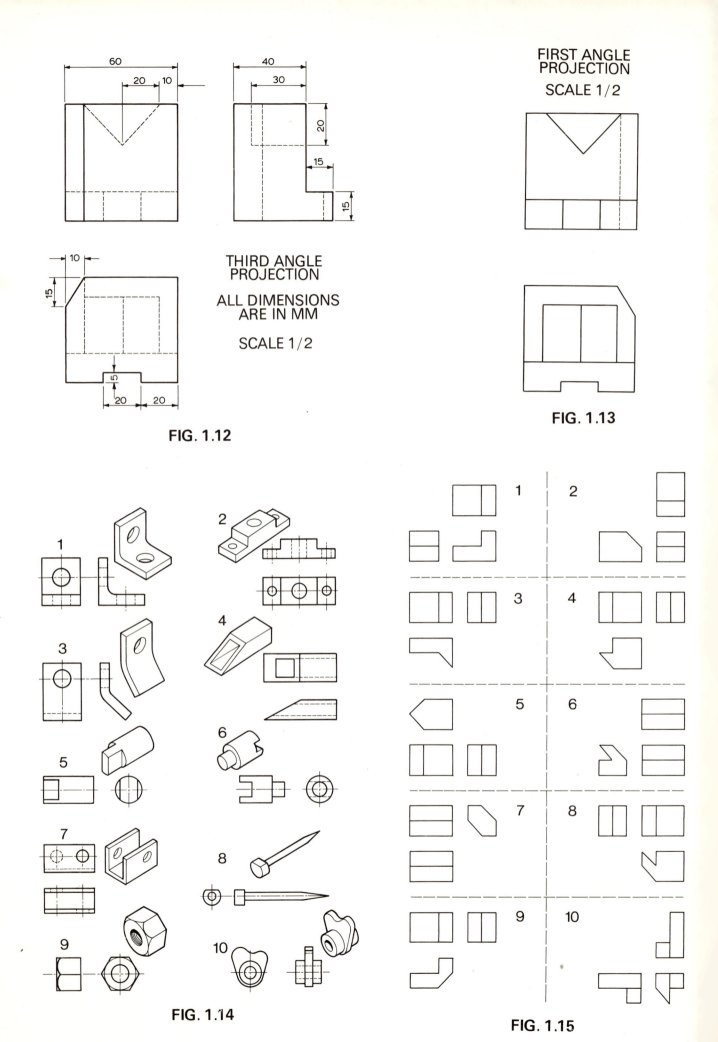

THIRD ANGLE
PROJECTION

ALL DIMENSIONS
ARE IN MM

SCALE 1/2

FIG. 1.12

FIRST ANGLE
PROJECTION

SCALE 1/2

FIG. 1.13

FIG. 1.14

FIG. 1.15

EXERCISE 1.3 Conversion from Third to First Angle Projection

Figure 1.12 shows three views of a Vee block. The views are badly chosen and there is a great deal of hidden detail. The object is to be redrawn in first angle projection, and it should be possible to find some more suitable views. Study Fig.1.12 carefully and get to know the object. Make a simple sketch if you find that a pictorial view helps you. Decide on the number of views and the viewing directions for a clearer drawing.

Make the drawing on a piece of graph paper and use first angle projection. As explained above, first and third angle projection are exact opposites and you may find it helpful to make a sketch of the views that you propose to draw, in order to check that they are correctly positioned. When you have completed the views, add the necessary labels and then compare your drawing with *Figure 1.13*.

Summary The Vee block of Exercise 1.3 is a simple object and two views are sufficient to show all its features. The two left-hand views in Fig.1.12 are poorly chosen because much of the detail is hidden behind the visible surfaces. The opposite viewing directions result in no hidden details and these views have been chosen for Fig.1.13. The lower view is what would be seen from the top of the upper view, so the drawing is correctly arranged for FIRST angle projection. If a view from the right-hand side were to be drawn, it should be placed on the left.

EXERCISE 1.4

Figure 1.14 shows a pictorial view and orthographic drawings of various objects. State whether first or third angle projection has been used in the orthographic drawing. Answers are at the end of this Chapter.

EXERCISE 1.5

Fig.1.15 shows orthographic drawings of various objects. State whether first or third angle projection has been used. N.B. Some of the arrangements are ambiguous. Answers are at the end of Chapter.

EXERCISE 1.6

Figure 1.16 shows pictorial views of various objects. Select views that would most clearly show the object and show how these views should be arranged in both first and third angle projection.

Answers to Exercise 1.4

1. 1st;   2. 1st;   3. 3rd;   4. 3rd;
5. 1st;   6. 3rd;   7. 1st;   8. 1st;
9. 1st;   10. 3rd.

Answers to Exercise 1.5

1. 3rd;   2. 3rd;   3. 1st;   4. 1st;
5. 1st;   6. 3rd;   7. Both; 8. 3rd;
9. Both; 10. 1st.

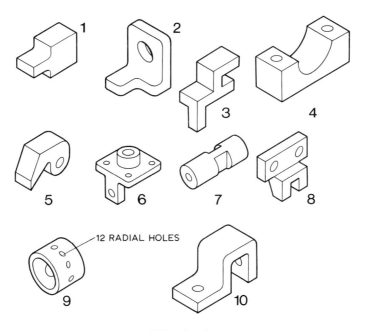

FIG. 1.16

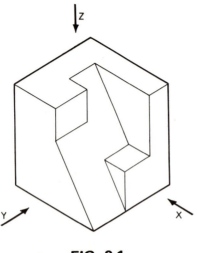

FIG. 2.1

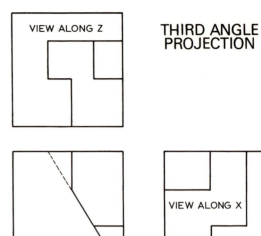

THIRD ANGLE
PROJECTION

VIEW ALONG Z

VIEW ALONG Y

VIEW ALONG X

FIG. 2.2

THIRD ANGLE PROJECTION

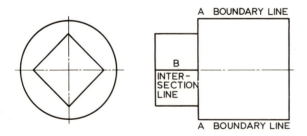

A   BOUNDARY LINE

B

INTER-
SECTION
LINE

A   BOUNDARY LINE

FIG. 2.3

THIRD ANGLE PROJECTION

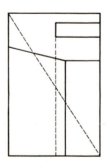

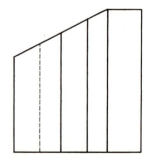

FIG. 2.4

CHAPTER 2

Visualization

In the previous Chapter, it was stated that the disadvantage of orthographic projection is that one has to acquire some skill in order to visualize the object. The purpose of this Chapter is to provide some practice at this activity so that future drawings may be read and understood as rapidly as possible.

Figures 2.1 and *2.2* show a pictorial drawing and an orthographic drawing of the same object. Fig.2.2 shows views as seen in the directions of arrows X, Y and Z of Fig.2.1.

It is not too difficult to visualize the separate views from a pictorial drawing of the object. However, until one has a certain degree of fluency in orthographic projection, it is not so easy to picture the solid object from the separate views. You can see this for yourself by turning to Fig.2.2 for a time and trying to picture the solid object, without reference to Fig.2.1.

It is important to remember that, in any orthographic drawing, a continuous thick line and a hidden detail (thin) line can indicate either:
  (1) the boundary between the material of which the object is made and the surrounding medium (which will be referred to as "air"), or
  (2) the intersection of two surfaces.

The distinction is not always clear and in Fig.2.2 nearly every line is both a boundary line and an intersection line. There is an example of a boundary line which is not also an intersection line in *Figure 2.3*. The lines labelled A do not represent the intersection of two surfaces: their only function is to show the boundary between the material and the air.

The line labelled B represents only the intersection of two surfaces. The test for distinguishing between the two types is that, if it were possible to grind and smooth the material at the line, the intersection line would be removed, whereas the boundary line would merely be shifted.

Sometimes, lines will be superimposed because of the viewing direction; what appears as a single line in one view may be two or more boundary or intersection lines overlaid. The other views of the drawing have to be consulted before the situation becomes clear. If a visible line is coincident with a hidden detail line then, naturally, the visible line takes precedence. This situation may be avoided by a careful choice of view on the part of the draughtsman, but it is something for which the reader of a drawing should be watching.

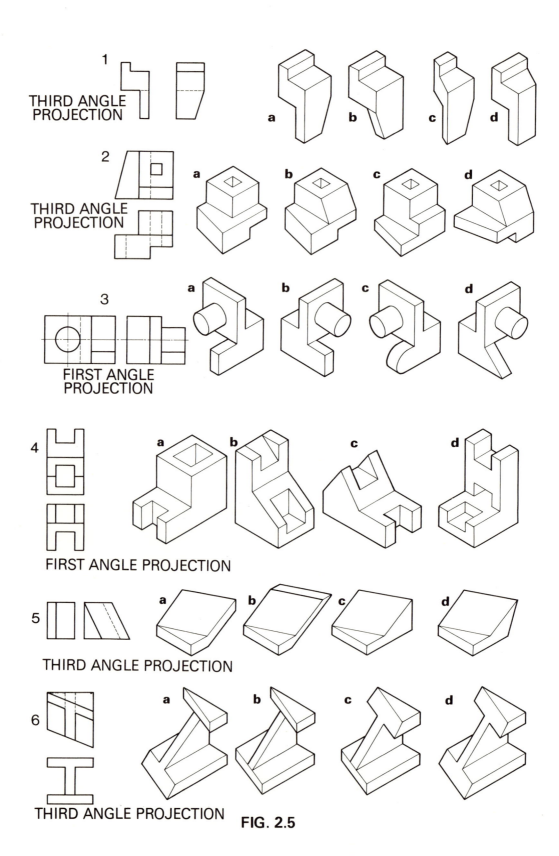

1

THIRD ANGLE PROJECTION

a   b   c   d

2

THIRD ANGLE PROJECTION

a   b   c   d

3

FIRST ANGLE PROJECTION

a   b   c   d

4

FIRST ANGLE PROJECTION

a   b   c   d

5

THIRD ANGLE PROJECTION

a   b   c   d

6

THIRD ANGLE PROJECTION

a   b   c   d

FIG. 2.5

The following steps are essential when reading any orthographic drawing for the first time:

(1) Identify the type of projection. The drawing will often be labelled FIRST ANGLE PROJECTION or THIRD ANGLE PROJECTION, or one of the symbols shown in B.S.308 may be used. In the absence of a definite specification, it may be possible to discover the type of projection by inspection, just as in the exercises at the end of the last Chapter. If the projection is ambiguous or if you are uncertain about the actual system, it is much better to ask than to make the wrong assumption.

(2) Identify the various views and make sure that you have understood their relationship to one another.

(3) Look at the views in more detail and identify the same feature in each view in which it appears. It may sometimes be necessary to use a straight-edge in order to see how the features line up between views. There are various unusual combinations of lines that occur with sufficient frequency to be worth special attention: they will be dealt with in the exercises below. Bear in mind that a rectangular shape can be many things, as well as a horizontal or vertical surface: it may be an inclined surface or it may be a curved surface.

EXERCISE 2.1

Figure 2.4 shows two views of a block. Go through the steps described above and try to get a clear idea of the solid object. Quite a lot of the information about the block is presented as hidden detail lines. Make a sketch of the object and check that it agrees with the two views in the figure.

When you have finished, compare your sketch with *Figure 2.6*. Notice how the various features produce the full and hidden detail lines of the orthographic drawing.

EXERCISE 2.2

Figure 2.5 shows orthographic drawings of various objects, each with a selection of pictorial views. Study the drawings and try to picture the solid object without looking at the pictorial views. When you have visualized the object, select the pictorial view which is correctly matched with the orthographic drawing.

When you have finished, turn to the answers at the end of this Chapter. You should pay special attention to any wrong answers to Fig.2.5. Most people find that planes normal to the viewing directions give very little trouble; planes that are inclined, but appear as a line in one of the views, give more trouble and planes that appear neither in their true shape, nor as a straight line, give most trouble at first.

Engineering practice and visual noise

Engineering components are rarely machined on all surfaces; only a few have no rounded edges at all. The majority of the objects shown in this Chapter should not be regarded as typical of engineering practice; they are only intended as exercises in visualization.

Dimensions can make a drawing more difficult to understand, although they are an essential part of the communication medium. For this reason, some of the exercises at the end of this Chapter have dimensions and notes added. These exercises provide practice at concentration, as it is important to ignore the "visual noise" and observe only the shape of the object.

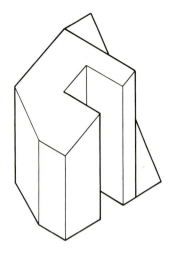

FIG. 2.6

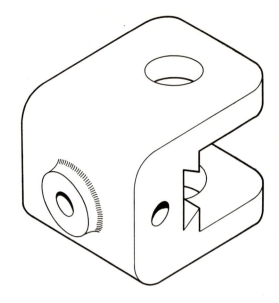

FIG. 2.9

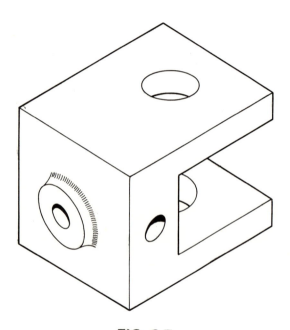

FIG. 2.7

FIRST ANGLE PROJECTION

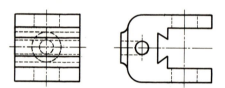

FIG. 2.10

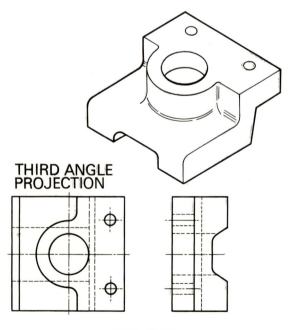

THIRD ANGLE
PROJECTION

FIRST ANGLE PROJECTION

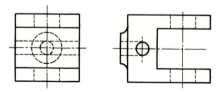

FIG. 2.8

FIG. 2.11

EXERCISE 2.3

Figure 2.7 is a pictorial view of a casting. Study it carefully and get to know the object. *Figure 2.8* shows two views (not to the same scale) of the same object; check that they agree with Fig.2.7.

The bracket is to be modified so that it fits on to a different shape of bar and its general appearance is improved. These modifications are shown in *Figure 2.9*. How should the views of Fig.2.8 be modified? Make a sketch of an orthographic drawing of the modified bracket.

When you have finished, compare your drawing with *Figure 2.10*.

Fig.2.10 shows two views of the bracket illustrated in Fig.2.9. Notice the differences between Figs. 2.8 and 2.10. Because of the viewing direc-

tions, not all the curves introduced in Fig.2.9 can be seen. There is a lot of hidden detail which would be very difficult to interpret without the pictorial view. The views shown do not fully define the bracket for it would be possible to interpret them as indicating square corners on the forked end. This ambiguity can usually be avoided by a careful choice of views.

EXERCISE 2.4

Figure 2.11 contains a pictorial view of a component and two views, in orthographic projection, of the same sort of object but not an identical one. Study the orthographic drawing and identify the differences between the two objects. By partially tracing the existing pictorial view, make a sketch of the modified component.

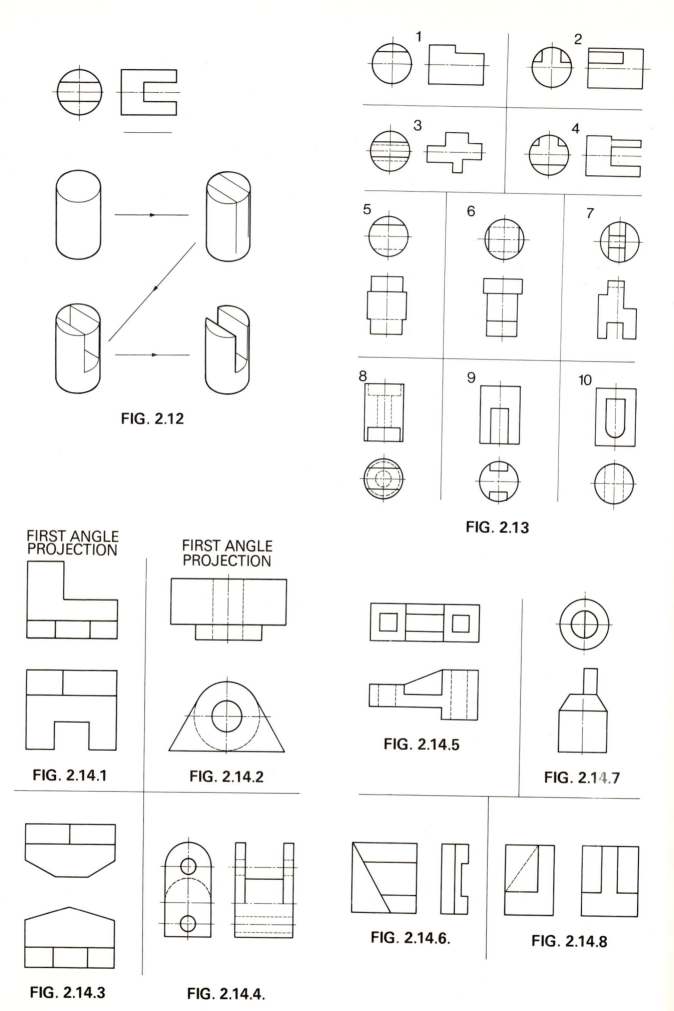

FIG. 2.12

FIG. 2.13

FIRST ANGLE
PROJECTION

FIRST ANGLE
PROJECTION

FIG. 2.14.1

FIG. 2.14.2

FIG. 2.14.5

FIG. 2.14.7

FIG. 2.14.3

FIG. 2.14.4.

FIG. 2.14.6.

FIG. 2.14.8

Intersection of planes and curved surfaces

This type of intersection occurs very frequently in engineering components, so many of which are turned on a lathe. *Figure 2.12* is an example of how to sketch a pictorial view of a basically cylindrical component. The orthographic drawing at the top may be found, by inspection, to be in first angle projection. The first stage in the preparation of a pictorial view is a sketch of the basic cylinder. Next, two parallel lines are drawn on the top surface and down the curved surface. In the third stage, the other visible lines are sketched in. Finally, the superfluous lines are removed. This type of construction can be used for all the objects in the next exercise.

EXERCISE 2.5

Figure 2.13 shows ten objects; each of them could be manufactured from a cylindrical piece of material, 20 mm diameter, 30 mm long. For each drawing, identify and state the type of projection used and make a sketch of a pictorial view of the object. When you have finished, check that your sketches agree with the original drawings. Answers, for the type of projection, are at the end of this Chapter.

EXERCISE 2.6

The final set of exercises in this chapter is graded and contains some of the visual noise mentioned above. Work through the sixteen objects in *Figures 2.14.1 to 2.14.16* by the usual method of first identifying the projection, then examining the relationship between the views, and then familiarizing yourself with the various features. Make a pictorial sketch of each of the sixteen objects, in order.

When you have finished, compare your solutions with *Figures 2.15.1 to 2.15.16*.

Answers to Exercise 2.2

1. a   2. b   3. **b** and d
4. c   5. d   6. c

Answers to Exercise 2.5

1. 1st;   2. 3rd;   3. 1st;   4. 1st;
5. 3rd;   6. 1st;   7. 3rd;   8. 3rd;
9. 3rd;   10. Both.

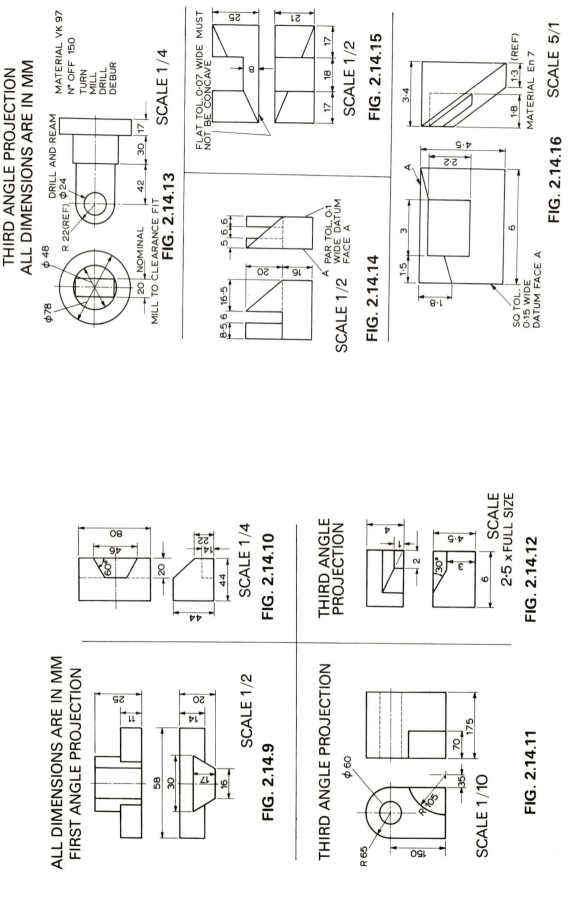

THIRD ANGLE PROJECTION
ALL DIMENSIONS ARE IN MM

MATERIAL VK 97
Nº OFF 150
TURN
MILL
DRILL
DEBUR

SCALE 1/4

FIG. 2.14.13

DRILL AND REAM ⌀24

R 22(REF)

⌀48

⌀78

20 NOMINAL

MILL TO CLEARANCE FIT

42

30

17

FLAT TOL.0·07 WIDE MUST NOT BE CONCAVE

25

21

17

18

17

8

SCALE 1/2

FIG. 2.14.15

PAR TOL.0·1 WIDE DATUM FACE A

5 6·6

5 6·6

8·5 6   16·5

20

16

SCALE 1/2

FIG. 2.14.14

3·4

1·3 (REF)

1·8

MATERIAL En 7   SCALE 5/1

4·5

2·2

3

1·5

1·8

6

A

SQ TOL. 0·15 WIDE DATUM FACE A

FIG. 2.14.16

80

46

60°

22

14

20

44

44

SCALE 1/4

FIG. 2.14.10

THIRD ANGLE PROJECTION

4

1

2

130°

4·5

3

6

SCALE 2·5 × FULL SIZE

FIG. 2.14.12

ALL DIMENSIONS ARE IN MM
FIRST ANGLE PROJECTION

25

11

20

14

58

30

17

16

SCALE 1/2

FIG. 2.14.9

THIRD ANGLE PROJECTION

⌀60

R 65

175

70

35

R 105

150

SCALE 1/10

FIG. 2.14.11

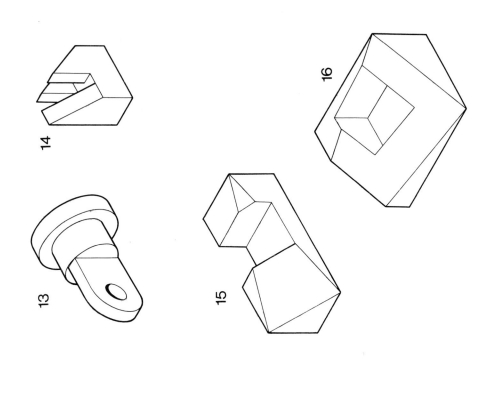

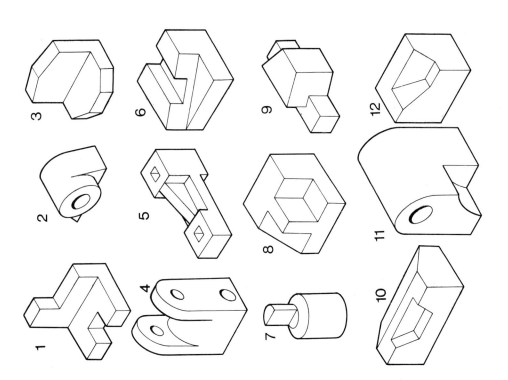

FIG. 2.15

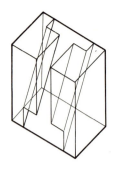

FIG. 3.1

FIG. 3.2

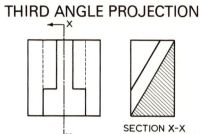

SECTION X-X

FIG. 3.3

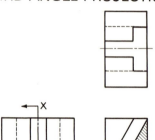

SECTION X-X

FIG. 3.4

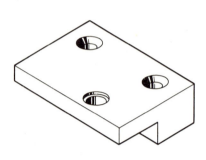

FIG. 3.5

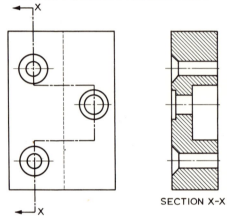

SECTION X-X

FIG. 3.6

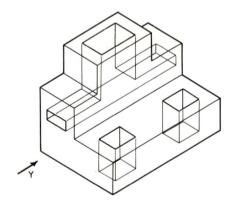

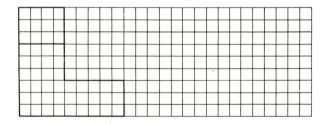

FIG. 3.7

CHAPTER 3

Sections

In the previous chapters, the only drawings that have been examined are those with views that show the external appearance of the object. Many engineering components are more complicated internally than they are externally. *Figure 3.1* is a sketch of a block made of some transparent material and all the edges are visible. Two views of this block are shown in third angle projection in *Figure 3.2*; here the material is opaque, like most practical materials are. The transparent material indicated in Fig.3.1 is a convenient technique for showing the internal details of the object. Notice how orthographic projection, with its use of visible and hidden detail lines, tends to make the object more difficult to visualize. Without the benefit of Fig.3.1, it would not be quite so easy to understand what all the hidden detail lines mean.

The internal details of the object may be shown by drawing a sectional view, normally labelled as a *Section*. In other words, that part of the object which conceals the details of interest is removed and only what remains is drawn. Normally, sections are drawn for planes cutting the object. *Figure 3.3* contains a section of the block that has been produced by removing everything to the right of the plane X-X and, in this way, one gets a clearer impression of the holes.

Notice how the section plane X-X is indicated. It is in accordance with the recommendations in British Standard 308 : 1964. The plane is shown as a thick chain line:

———— — ———— —— ————

complete with arrowheads and letters. The section is labelled accordingly. Where material has been cut by the section plane, this is indicated by means of continuous thin lines at, in this case, 45°; this evenly-spaced set of lines is called *section lining* or *hatching*.

Figure 3.4 shows three views of the same block. The sectional view is the link view (see Chapter 1). Notice that the upper view shows the complete block. It would be most confusing if the upper view were made a pedantic view on to the top of the section and showed only part of the object. All views that are not specifically chosen as sections should represent the complete object.

Many objects cannot be conveniently sectioned by taking only one section plane. It is necessary in such cases to consider a section on two or more planes. *Figure 3.5* shows such an object and *Figure 3.6* shows the same object cut on two parallel planes, with two discontinuities in the section line. Notice how, in the left-hand side of the section, the discontinuities in the section line are not shown; there is no way of avoiding showing them on the right-hand side.

EXERCISE 3.1

Figure 3.7 is a pictorial drawing of a block. The internal details have been indicated by thin lines, as though the object were made of a transparent material. In the squared grid is part of the view that would be seen in the direction of arrow Y. Complete this view using thin, hidden detail lines for the internal features, as for an opaque material. Choose a section that will show all the holes and project this section, from the existing view, to the right side of the grid. Complete the drawing by section-lining the sectional view and inserting the appropriate thick chain lines and labels.

23

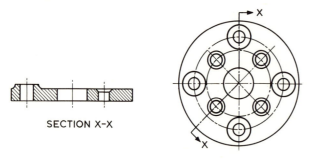

SECTION X-X

FIG. 3.8

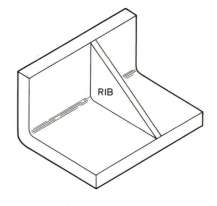

RIB

FIG. 3.13.1

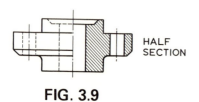

HALF SECTION

FIG. 3.9

THIRD ANGLE PROJECTION

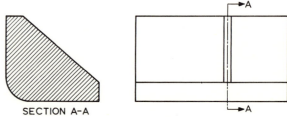

SECTION A-A

FIG. 3.13.2

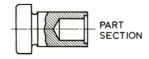

PART SECTION

FIG. 3.10

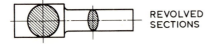

REVOLVED SECTIONS

FIG. 3.11

THIRD ANGLE PROJECTION

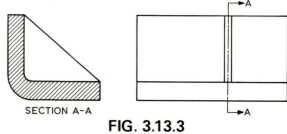

SECTION A-A

FIG. 3.13.3

REMOVED SECTIONS

FIG. 3.12

Other types of sectional views

Figure 3.8 shows, in the right-hand view, a circular object with four axes of symmetry. There are two basic types of hole, excepting the central one, and, in order to section through both types, the section planes are drawn as shown. As the section planes are not parallel, it would be impossible to project from both planes in the same direction. The sectional view on the left is not projected along either horizontal or vertical lines but is constructed independently of the other view.

Notice that the sectional view represents the two parts of the section in alignment, i.e. it would not actually be possible to produce such a view by cutting the solid object only along the section planes, unlike the section in Fig.3.6. Nevertheless, this sectioning device is so useful that it has been accepted as a recommendation in B.S.308. Part of the sectional view may be projected from the adjacent view, if so desired.

Figure 3.9 shows a *half section*. This is another useful method and is often used for objects with an axis of symmetry. There is no need to have the hidden detail lines on the left-hand side but they have been inserted in order to emphasize the symmetry in this example.

Figure 3.10 is a *part section*. There would be no point in sectioning the whole of such an object and a part section is taken to show the hole. The boundary of the section is shown by an irregular thick boundary line; this is the sixth type of line for engineering drawings which has been introduced in the book. (What are the other five?)

Figure 3.11 has two *revolved sections*. This is a convenient method of showing the shape of varying cross-sections. The section planes are indicated with <u>thin</u> chain lines and the sectional views are revolved about these lines. The outline of the section should be <u>thin</u>, to distinguish it from the remainder of the view.

Figure 3.12 has two *removed sections*. These are similar to revolved sections but are used when it is difficult to superimpose the cross-section on the existing view. The section planes are indicated with extended <u>thin</u> chain lines and the sectional views are revolved about these lines, outside the existing view. The outlines of the sections are <u>thick</u>.

Ribs

A rib is any thin, flat part of a casting or moulding incorporated in order to give the object greater strength and rigidity. A triangular rib is shown in *Figure 3.13.1*; the object is not typical of engineering practice and serves only to illustrate the sectioning of a rib. Two views of the object are shown in orthographic projection in *Figure 3.13.2*; a section plane is indicated which passes LONGITUDINALLY through the rib. Notice the result when the sectional view is pedantically drawn. The impression of solidness and bulk is quite unmistakable but completely misleading.

It is better to ignore the fact that the section plane passes through the rib and to draw the rib in outside view, as in *Figure 3.13.3*. This is the B.S.308 recommendation for ribs; other examples of this important exception to the principles of sectioning will be found below.

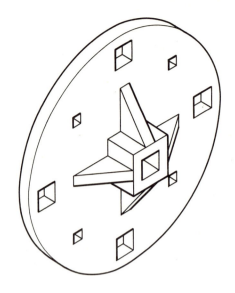

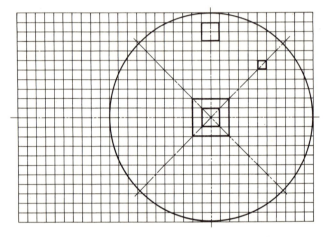

FIG. 3.14

THIRD ANGLE PROJECTION

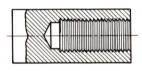

FIG. 3.16.1

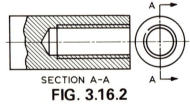

SECTION A-A

FIG. 3.16.2

THIRD ANGLE PROJECTION

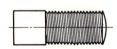

FIG. 3.17.1

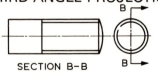

SECTION B-B

FIG. 3.17.2

EXERCISE 3.2

Study the pictorial view shown in *Figure 3.14*. Part of one view of the object is drawn on the grid below; complete this view and consider how best to section the object in order to show both sizes of non-central hole. Draw the sectional view in the grid provided; add all necessary section lines, section-lining and labels. When you have finished, compare your solution with *Figure 3.15*.

Fig.3.15 shows one reasonable method of sectioning the object. It is similar to the method shown in Fig.3.8, but here the top part of the section has been projected, horizontally, from the adjacent view. The lower part has been drawn by transferring radial distances, measured along the section line, to the appropriate vertical position in the sectional view. Notice how the rib has not been sectioned in the upper half and how it is visible beyond the section plane in the lower half.

An alternative method of sectioning this object would be to use two parallel planes; one passing vertically through the larger hole and the other passing through the smaller hole. The discontinuity could be at any convenient place.

Exceptions and conventions

In the same way that a rib is not shown in section when the section plane passes longitudinally through it, so other common parts are treated in the same way. It would be pointless and misleading to section shafts, bolts, nuts, rods, rivets, keys, pins, shims and washers. Such components will normally appear with a section plane longitudinally through them in General Arrangement (Assembly) drawings only: they are illustrated in Chapter 10. However, two threaded features are so important that they will be treated here.

The *tapped hole*, or *internal screw thread*, is shown in *Figure 3.16.1*. The part section shows a blind tapped hole with the end of the hole made by the tapping size drill and the helices cut by the tap, in quite accurate detail. The cost of producing such a drawing would be prohibitive, so the convention shown in *Figure 3.16.2* is recommended. The crests and roots of the thread are represented by parallel lines and the section lining is taken as far as the tapping size hole. The end view of the hole is represented as shown, with the outer circle broken, to indicate the thread.

The *external screw thread* of a bolt, stud or screw is shown realistically in *Figure 3.17.1*. Again, the cost of the drawing would be prohibitive and *Figure 3.17.2* shows the recommended convention. The end view of this fastener is represented by two concentric circles but, in this case, the inner one is broken.

One way of remembering the convention is that the circle which would appear First in the process of manufacture (the tapping hole or the plain bar) is shown as the Full circle.

In Fig.3.17.2, the section plane passes through the fastener but it is not sectioned; plane B-B would normally be drawn so as to reveal the internal details of an assembly of parts.

Special section lining

When a large area of a section has to be shown, the recommendation of B.S. 308 is that the edges alone should be section-lined. Such an area is shown in *Figure 3.18*.

Normally, section lining is at 45° to the edges of the paper and spaced in relation to the area to be covered. However, there are cases, as in *Figure 3.19*, when 45° lining is inappropriate and another angle is used. The sense and spacing of section-lining can be varied in order to distinguish between adjacent parts of an assembly.

EXERCISE 3.3

Describe the errors or omissions in the drawings shown in *Figures 3.20 to 3.23*.

THIRD ANGLE PROJECTION

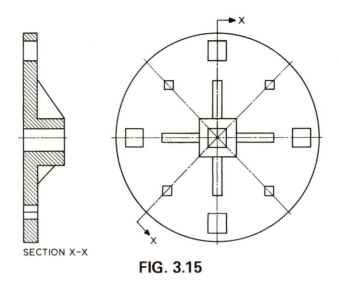

SECTION X-X

FIG. 3.15

THIRD ANGLE PROJECTION

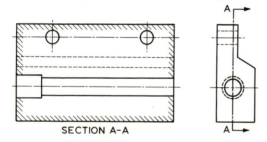

SECTION A-A

FIG. 3.18

THIRD ANGLE PROJECTION

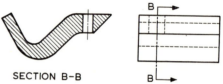

SECTION B-B

FIG. 3.19

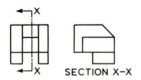

SECTION X-X

FIG. 3.20

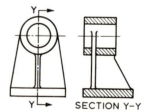

SECTION Y-Y

FIG. 3.21

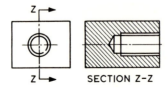

SECTION Z-Z

FIG. 3.22

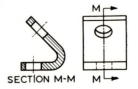

SECTION M-M

FIG. 3.23

CHAPTER 4

The Use of Drawing Instruments and Equipment

This Chapter is not intended to be a description of Drawing Office Practice in the use of draughting machines and other equipment; it is a summary of those instruments a student is most likely to use. Drawing equipment will be considered under five headings: the material on which the drawings are made, the methods of supporting that material, methods of marking the material, methods of removing marks from the material, and miscellaneous aids.

Materials

Most engineering drawings are made in pencil on translucent materials to facilitate the rapid preparation of prints. Translucent paper can tear all too readily in inexperienced hands and, especially if prints are not required, it is convenient to use opaque drawing paper for some exercises. The paper should be of sufficiently high quality to withstand some rough handling and the erasure of the inevitable mistakes.

Translucent cloth and plastic drawing film are other very useful materials from which prints can be made. Many companies have standard sizes of drawing sheets with the company's name and other titles and headings printed on the sheet so that they are reproduced during the copying process.

The International Standards Organization's A-series of drawing sheet sizes is based on a rectangle of 1 square metre area, the sides of which are in the ratio $1:\sqrt{2}$. The sizes are obtained by dividing the next larger size into two equal parts, the division being parallel to the shorter side. The A-series has the following dimensions:

Designation	Millimetres
A0	841 × 1 189
A1	594 × 841
A2	420 × 594
A3	297 × 420
A4	210 × 297

Supporting the drawing material

It is important that the paper or other material is supported on a flat, dimensionally stable surface which is not so hard that there is no "give" when pressure is applied.

FIG. 4.1

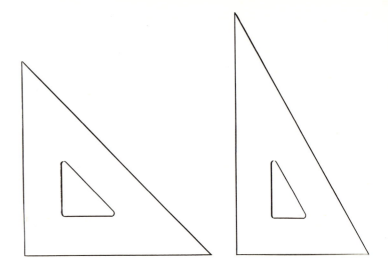

FIG. 4.2

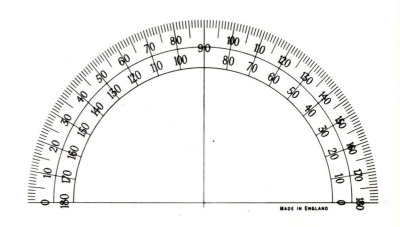

FIG. 4.4

FIG. 4.3

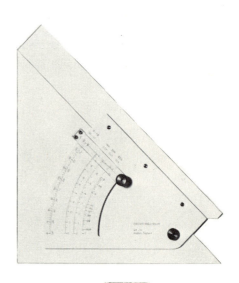

FIG. 4.5

Nearly all drawing boards contain wood and their cost reflects the quality of the timber. The cheapest are made of plywood but these are only suitable for low-quality work and they have a relatively short useful life. Better quality boards are constructed from blockboard with a wood veneer or plastic surface. The best boards are constructed from well-seasoned timber and have battens on the underside. If a tee-square is to be used, a hard insert is let into the working edge. A typical good quality board for student use is shown in *Figure 4.1*.

The paper, on which the drawing is to be made, may be laid directly on to the plastic-surfaced type of board, or on to a paper backing-sheet if the board has a wooden surface. There must always be some resilience under the drawing material. The paper is held in place with draughting tape or with spring clips, like the one shown in *Figure 4.2*. Drawing pins can ruin any board.

Drawing boards are, of course, slightly larger than the corresponding A-series paper sizes:

Designation	Board Size Millimetres
AO	950 × 1 270
A1	650 × 920
A2	470 × 650
A3	336 × 470

The board may be used on a suitable table, propped up at a convenient angle. The larger boards are often mounted on a positioning mechanism and provided with a parallel motion straight edge. Such draughting machines are very expensive and their use can usually be justified only by the productivity of the draughtsman who would be employing one continually.. A typical, medium size draughting machine is shown in *Figure 4.3*.

For the board shown in Fig.4.1, horizontal lines are drawn with the aid of a tee-square. The stock of the tee-square must always be in line contact with the edge of the board. Vertical lines are drawn by employing a set-square, resting on the tee-square. For certain angles, the sloping edge of the set-square is used. To cover all possible angles, the three instruments illustrated in *Figure 4.4* are required: they are, from left to right, a 60º set-square, a 45º set-square, and a protractor.

An adjustable set-square is often useful. With one of these it is possible to dispense with the 45º set-square and the protractor, but not entirely with the 60º set-square. *Figure 4.5* illustrates a robust adjustable set-square.

Marking the drawing material

Most drawings are made with pencil and the leading manufacturers produce ranges of about twenty degrees of hardness. For drawings on translucent materials, it is generally found that draughtsmen use mainly 2H grade for line work and F grade for lettering. For students' drawings on opaque paper, the most suitable grades are HB for lettering and H and 2H for line work. Sometimes, it may be convenient to use a 3H grade for certain construction lines.

The mechanical type of "lead" holder shown in *Figure 4.6* is very useful. The "lead" is firmly held in the metal jaws but it can be quickly released for changing or resharpening.

"Leads" are sharpened to give either a conical point, as in *Figure 4.7* or a chisel point, as in *Figure 4.8*. After removing sufficient wood from the pencil with a sharp blade, the point is prepared by rubbing the "lead" on a piece of fine glass paper. Blocks of about ten sheets are available for this purpose. The conical point is used for lettering and dimensioning and the chisel point, which wears less rapidly, is ideal for line work.

FIG. 4.6

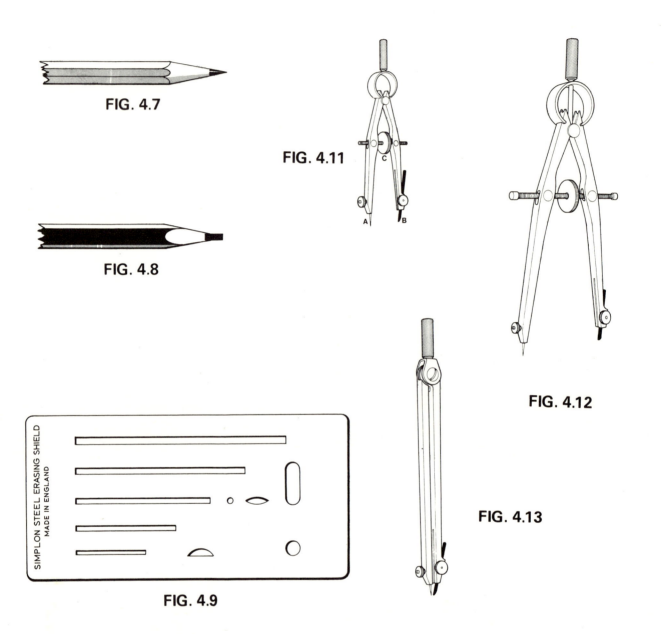

FIG. 4.7

FIG. 4.11

FIG. 4.8

FIG. 4.12

FIG. 4.9

FIG. 4.13

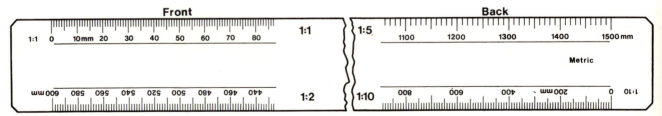

FIG. 4.10

Occasionally, it is necessary to use drawing ink for illustrations, charts, graphs and some special engineering drawings. The need to use ink is far less than it was before print making machines reached their present standard of performance. Ink drawing should be attempted only when a certain skill with pencil drawings has been acquired. Special pens are required for ink lines but some pencil compasses may be fitted with a pen head.

Removing marks

Pencil marks may be removed from most types of drawing materials with a good quality eraser which is effective without damaging the paper. There are various types on the market; some are encased in wood and can be sharpened, like a pencil, so that their action is restricted to a small area of the paper. Alternatively, a thin sheet-metal erasing shield, as in *Figure 4.9*, may be used to prevent the erasure of correct lines adjacent to mistakes.

Tracing erasers will remove both pencil marks and dry drawing ink but it is all too easy to damage the paper. If the printing process to be used makes use of reflected light, it is possible to "paint out" mistakes with the compound "Snopake" and to draw on top of the dry painted area.

A soft "art gum" or "art cleaner" block is very useful for cleaning finished drawings. Eraser dirt should be removed by flicking with a clean stockinette duster, which can also be used for wiping the tee-square and set-squares.

Other drawing instruments

Measurements are made using a scale like the one shown in *Figure 4.10*. It should have an oval cross-section and sets of millimetre graduations. Modern plastic scales are far more robust than the traditional box-wood type.

Circles and circular arcs, of up to about 25 mm radius, are drawn using the type of spring bows illustrated in *Figure 4.11*. Notice that they have a removable shouldered point (A), a replaceable "lead" (B) and a knurled wheel (C) for setting the radius. The "lead" should be sharpened on one side only.

Radii should never be set by sticking the point into the scale. Instead, mark the required radius on an odd piece of paper, stick the point in one mark, and adjust the knurled wheel until the "lead" draws through the other mark.

For larger radii, of up to about 120 mm, spring compasses of the type shown in *Figure 4.12* are ideal for students. When fitted with an extension bar they may be used for radii of up to 240 mm. The traditional design of draughtsman's compasses, shown in *Figure 4.13*, will also draw these larger radius arcs but there is no positive setting of the radius. They too may be fitted with an extension bar.

Very small radius arcs (1 - 12 mm radius) may be drawn very quickly, without having to mark the position of the centre, with the aid of a template. Two types are illustrated in *Figure 4.14*.

For curves other than circular arcs, a set of french curves or, at least, one similar to that shown in *Figure 4.15* is very useful. The flexible curve shown in *Figure 4.16* is a fairly good substitute.

At the drawing board

It is well worth developing the ability to produce neat, attractive drawings, even if this is not the means by which one earns one's livelihood. The following suggestions are made with that aim in view:

(1) Always plan the arrangement of the whole drawing on the paper before making any marks whatsoever.
(2) Build up the drawing from centre lines and base lines which are common to two views.
(3) Do not try to complete one view before proceeding to another. It is much better to draw in parts of each view which are aligned along horizontal or vertical lines and work on each of the views in some sort of rotation.
(4) Keep your pencils sharp so that your lines are always crisp. Use the chisel point as much as possible.
(5) Practice, from the start, the maintenance of a clear distinction between the two thicknesses of lines. Thick lines, whether straight or curved, should be between two and three times thicker than thin lines.
(6) Keep all your instruments clean. Remove all loose matter from your drawing with a duster, not with your hand.

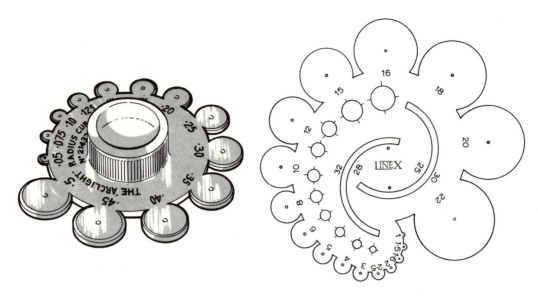

FIG. 4.14

FIG. 4.15

FIG. 4.16

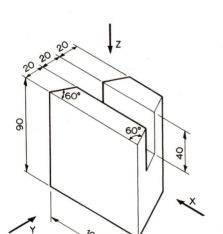

ALL DIMENSIONS ARE IN MM

FIG. 4.17

ABCDEFGHIJKLMNOPQRSTUVWXYZ
0123456789&

FIG. 4.18

EXERCISE 4.1

Figure 4.17 is a pictorial drawing of a block. Draw, scale 1/1, in third angle projection, views as seen in the directions of arrows X, Y and Z. Plan the layout and spacing of the drawing before you begin it. Insert centre lines and hidden detail lines but do not add dimensions.

When the views are completed, practise freehand printing of the style shown in *Figure 4.18* by adding the labels THIRD ANGLE PROJECTION and SCALE 1/1. Also print a title and your name. Use very thin horizontal lines as guide lines to make your printing about 3 mm high.

The use of springbows and compasses

Circular arcs should always be drawn in before the straight lines that are tangential to them. It is very much easier to blend the two together if they are drawn in this order. When drawing the arc, the centre should be located and marked first, then the radius should be set by making two marks on an unwanted piece of paper, sticking the point in one mark and adjusting the instrument so that the "lead" draws

through the other mark.

Construction lines required for the location of the centre of the arc should be so thin that they do not have to be erased.

EXERCISE 4.2

Figure 4.19 shows two views of a component in first angle projection. Draw, scale 1/1 and in third angle projection, the following views:

(a) the existing left-hand view and
(b) a sectional view on A-A.

Add all the necessary section-lining, centre lines and section line, labels and titles. Do not add dimensions. Check that you have a consistent difference between the thick lines and the thin lines.

Arc blending

It is easy enough to blend an arc with a straight line but much more difficult to blend two arcs having a common tangent. *Figure 4.20* illustrates this type of problem. (The tiny object looks rather unusual when drawn to such a scale. Cultivate the habit of immediately looking at the scale when you see a drawing for the first time.)

FIRST ANGLE PROJECTION

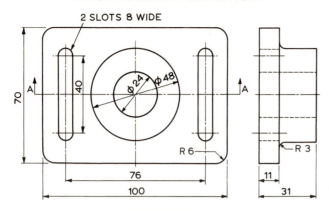

ALL DIMENSIONS ARE IN MM
SCALE 1/2

FIG. 4.19

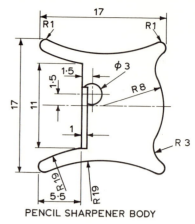

PENCIL SHARPENER BODY

ALL DIMENSIONS ARE IN MM
SCALE 2/1

FIG. 4.20

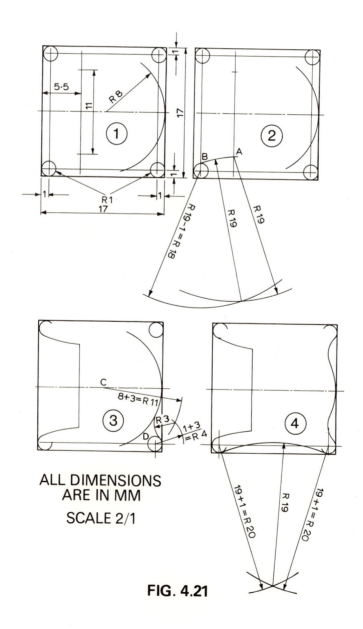

ALL DIMENSIONS
ARE IN MM

SCALE 2/1

FIG. 4.21

Figure 4.21 illustrates the stages of constructing the outline of the sharpener. Construction lines which are no longer required have been omitted from successive diagrams.

In stage (1) of Fig.4.21, the basic 17 mm square has been drawn and inside it are four lines parallel to the sides and 1 mm away. The intersections of these lines fixed the centres of the 1 mm radius arcs. It is a straightforward matter to construct the 11 mm line and the 8 mm radius arc.

Stage (2) shows how the inside 19 mm arc is constructed. The arc must pass through point A and therefore its centre lies on an arc, centre A, radius 19 mm. The arc in question and the lower left 1 mm radius circle must have a common normal and so the centre of the arc must lie on an arc, centre B, radius 19 − 1 = 18 mm. Where the two arcs intersect is the centre of the necessary 19 mm radius arc.

Stage (3) shows how the 3 mm arc is constructed. This arc has common normals with both the 8 mm arc and the lower right 1 mm arc. Where two arcs, one centre C, radius 8 + 3 = 11 mm, and the other centre D, radius 1 + 3 = 4 mm, intersect is the centre of the necessary 3 mm radius arc.

The centre of the other 19 mm arc is found in exactly the same way as the centre of the 3 mm arc. This has been done in stage (4).

The constructions described above occur quite often in engineering drawings and are worth remembering.

EXERCISE 4.3

Without reference to Fig.4.21 or the text relating to it, draw out the outline shown in Fig.4.20 using a scale of 5/1. Be careful to keep the lines and the arcs uniformly thick.

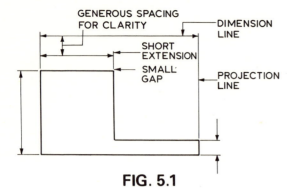

GENEROUS SPACING FOR CLARITY

DIMENSION LINE

SHORT EXTENSION

SMALL GAP

PROJECTION LINE

FIG. 5.1

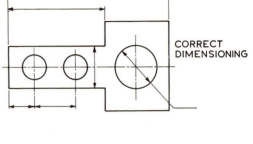
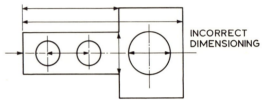

CORRECT DIMENSIONING

INCORRECT DIMENSIONING

FIG. 5.2

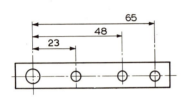

65

48

23

FIG. 5.3

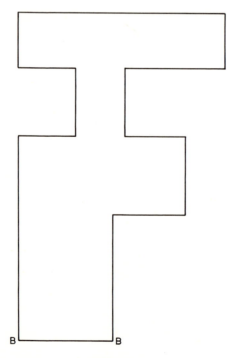

B   B

FIG. 5.4

CHAPTER 5

Dimensioning

The purpose of an engineering drawing is to enable a component to be manufactured or a set of components to be correctly assembled. In the former case, it is necessary to define the component completely and unambiguously; it is essential to state every measurement required in the clearest possible way. Furthermore, the way in which the dimensioning is done should take account of the methods by which the object is to be manufactured. In this Chapter, these two aspects of dimensioning will be considered separately, as far as is possible.

General principles

The dimension is stated next to a thin dimension line which has arrowheads at either end. The arrowheads touch either the appropriate features of the object or thin projection lines extended from these features. *Figure 5.1* shows the recommendations of B.S. 308.

Projection lines enable the dimensions to be placed outside the L-shaped outline. The projection lines start just clear of the outline and extend just beyond the arrowhead. Parallel dimension lines are well spaced for clarity. Dimension lines may be placed within the outline, if there is some good reason for doing so. Arrowheads should be slim and have a minimum length of about 3 mm.

Figure 5.2 shows the correct and incorrect ways of using projection and centre lines for dimensioning. The incorrect dimensioning is manifestly lacking in clarity.

Where several dimensions are stated from a common datum line, the method shown in *Figure 5.3* is recommended. Notice that the dimension figures are placed near the appropriate arrowhead.

EXERCISE 5.1

Draw a complete, clear set of projection and extension lines for the outline shown in *Figure 5.4*. Do not add any figures alongside the dimension lines. All vertical dimensions are to be taken from a common datum line through B-B.

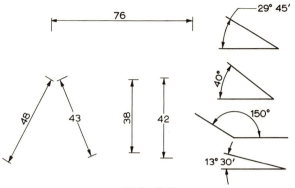

FIG. 5.5

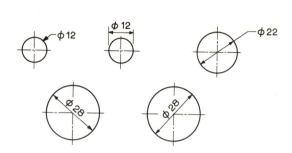

FIG. 5.10

FIG. 5.6

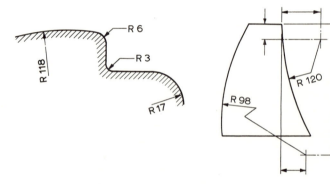

FIG. 5.11

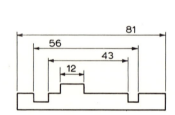

FIG. 5.7

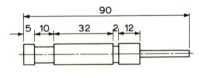

FIG. 5.8

THIRD ANGLE PROJECTION

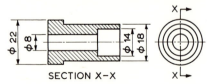

SECTION X–X

FIG. 5.9

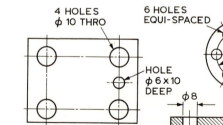

4 HOLES
φ 10 THRO

6 HOLES
EQUI-SPACED

HOLE
φ 6 x 10
DEEP

FIG. 5.12

Arrangement of dimensions

Figures 5.5 to 5.9 show various methods of dimensioning: each one illustrates some important point or points.

In *Figure 5.5*, the numbers are orientated so that they may be read from either the bottom or the right-hand edge of the drawing.

Figure 5.6 shows how narrow spaces are dimensioned by using inward pointing arrowheads.

In *Figure 5.7*, notice how the parallel dimension lines have the dimension figures staggered for clarity.

Figure 5.8 illustrates the advantage of placing overall dimensions outside intermediate dimensions: if the intermediate ones were outside, there would be a confusing crossing of lines.

Dimensions of diameters should be placed on the most appropriate view, to ensure clarity. The set of concentric circles in the right-hand view of *Figure 5.9* is difficult to understand but the sectional view makes it clear what each circle means. The dimensions are better placed on the sectional view. B.S.308 recommends that a diameter is indicated by, for example, ϕ 22 (as in the figure) but 22 DIA is used by many draughtsmen.

Circles, radii and holes

Circles are dimensioned by one of the methods shown in *Figure 5.10*, while *Figure 5.11* shows that the radii of arcs are dimensioned by a line lying along a radius. If the centre of the arc need not be located, it need not be indicated. If the centre must be located but lies in an inconvenient place, a fictitious centre may be created and the true position specified; the part of the dimension line that bears the arrowhead is along a true radius.

Holes are dimensioned by one of the methods shown in *Figure 5.12*. Identical holes may be specified by a leader line that points to only one of the holes but specifies the number of such holes. When the depth of a hole is stated, but not drawn, the depth refers to the cylindrical part of the hole only. The conical end formed by the 118° (or thereabouts) angle at the tip of a drill is not included in the depth. Holes whose centres lie on the circumference of a circle and subtend equal angles at the centre may be specified for position in the way which is illustrated.

THIRD ANGLE PROJECTION
SCALE 1/2

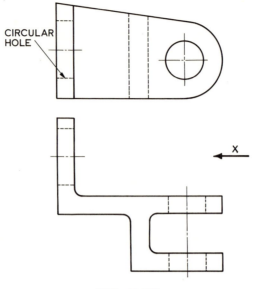

CIRCULAR
HOLE

X

FIG. 5.13

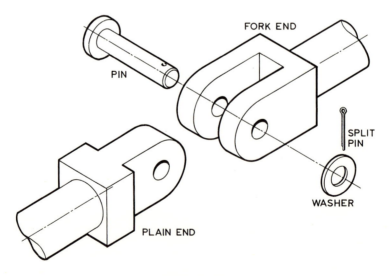

FORK END

PIN

SPLIT
PIN

WASHER

PLAIN END

FIG. 5.14

EXERCISE 5.2

Two views of a bracket are given in *Figure 5.13*. By scaling from the drawing (a practice to be deplored but which is necessary for this exercise) draw out the given views and a view as seen in the direction of arrow X, in FIRST angle projection. Allow ample space between the views. Add all necessary dimensions and labels.

Choose the dimensions carefully and use only the methods described in this Chapter. Always ask yourself: is this the right view on which to state this dimension? All dimensions should be stated in millimetres and a note to this effect should be inserted on the drawing.

Dimensioning for manu-facture and function

Figure 5.14 is a pictorial drawing of an exploded collection of parts which assemble to form a simple rotating joint between two rods of circular cross-section. Only part of the rod is shown on each end. The five components are assembled by inserting the plain end into the fork end so that the holes are aligned, pushing the pin through the holes, placing the washer on the end of the pin, and dropping the split pin through the hole in the pin. The assembly is kept together by opening up the split pin, in the usual way.

Figures 5.15.1 and *5.15.2* show two alternative ways of dimensioning the plain end and *Figures 5.16.1* and *5.16.2* show two alternative ways of dimensioning the fork end. (N.B. The designs of the components do not represent good engineering practice because there are too many sharp edges; they have been simplified in order to reduce the number of dimensions in the figures.) Both sets of dimensions would enable the ends to be manufactured as each defines the geometry in a logical and practical form.

THIRD ANGLE PROJECTION
ALL DIMENSIONS ARE IN MM
SCALE 1/2

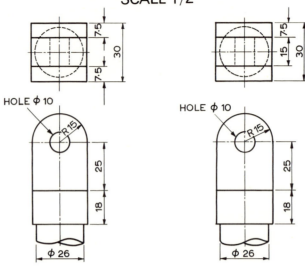

FIG. 5.15.1   FIG. 5.15.2

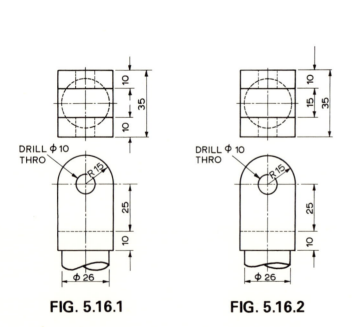

FIG. 5.16.1   FIG. 5.16.2

FIG. 5.17.1

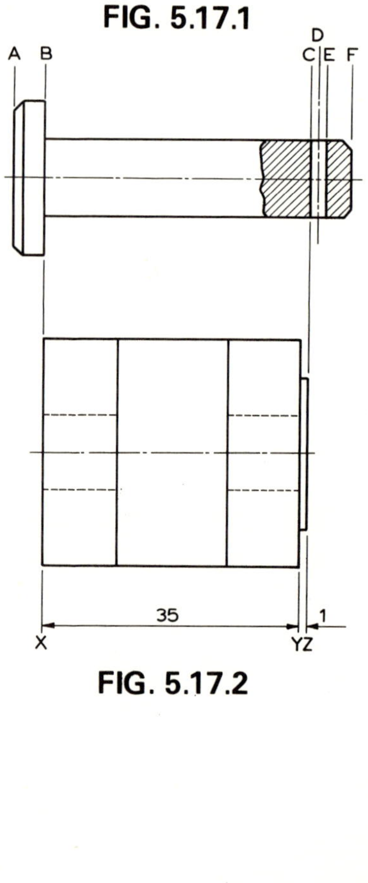

FIG. 5.17.2

Now consider the effect of a small error on some of the sizes. For convenience, it will be assumed that each dimension can be held to a tolerance of ± 0·1 mm. This would mean that, in Fig. 5.15.1, the worst thing which could happen is for the 30 mm dimension to be 0·1 mm oversize and both of the 7·5 dimensions to be 0·1 undersize. The thickness of the plain end would then be

$$(30 + 0·1) - 2(7·5 - 0·1) = 15·3 \text{ mm}$$

Turning now to Fig.5.16.1, the 35 mm dimension will be 0·1 mm undersize and both of the 10 mm dimensions will be 0·1 mm oversize. The width of the slot in the fork end would then be

$$(35 - 0·1) - 2(10 + 0·1) = 14·7 \text{ mm}$$

The net result is that the plain end is 0·6 mm thicker than the slot into which it is supposed to fit (and be free to rotate) and the assembly will not function as intended. Given the tolerances of ± 0·1 mm on all dimensions, it is equally possible that the plain end would be 0·6 mm thinner than the slot, but the cumulative error is much larger than it need be.

Now consider Figs. 5.15.2 and 5.16.2. The worst condition is that the plain end is 0·1 mm oversize and the slot is 0·1 mm oversize. The net result is that the interference, the excess of material size over hole size, is 0·2 mm, or one third of the previous case.

Clearly, the dimensions used in Figs. 5.15.2 and 5.16.2 take account of the functioning of the assembly, as well as the manufacture of the individual components, and are preferable to the other method of defining the geometry.

As a general guide, components should be dimensioned so that mating sizes are specified directly and not indirectly. This may not always be possible (as the next example illustrates) but it is usually possible to minimize the cumulative error. The components considered above could always be assembled if the width of the plain end were dimensioned as 14·9 ± 0·1 and the width of the slot as 15·1 ± 0·1. At one extreme there would be a clearance space of 0·4 mm and, at the other, a perfect fit.

Now consider the assembly of the pin, fork end, washer and split pin (the plain end does not affect this arrangement). The pin is shown in *Figure 5.17.1*: how should it be dimensioned so as to minimize the cumulative error?

In this case the mating size is BC but this dimension should not be stated as it would make the manufacturing process the subject of some arithmetic. The hole can only be drilled by first locating its centre, and there is always a chance of a mistake when calculating BC + ½CE.

Figure 5.17.2 shows how the two mating distances are composed. The cumulative error will be minimized by stating the dimensions as shown. One other point should be noticed: it would be possible to manufacture sizes XY and YZ very accurately. Starting from an excess of material, it is a simple operation to remove a small amount from a flat surface. On the other hand, the hole in the pin requires some accurate positioning of the drill and, once the hole has been made, no correction is possible. As the mating dimension is difficult to produce when it is known in terms of two specified dimensions, there is all the more reason to express it as directly as possible.

ALL DIMENSIONS ARE IN MM

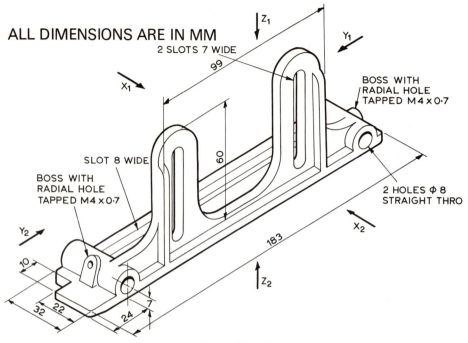

FIG. 5.18

Note. The metric thread designation M4 × 0·7 means:
International Standards Organization metric thread
with a nominal diameter of 4 mm and an axial pitch
of 0·7 mm.

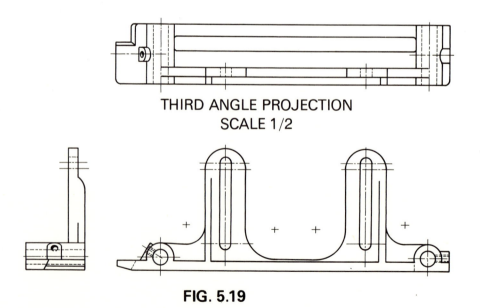

THIRD ANGLE PROJECTION
SCALE 1/2

FIG. 5.19

Figure 5.18 is a pictorial view of a depth-setting component for a circular saw. Many of the dimensions are missing. You are asked to prepare a dimensioned orthographic drawing of the object. Which views would most clearly show the object?

The component is fixed to the saw housing by means of two screws passing through the 7 mm wide slots; when these screws are tightened the depth of the cut is set. The two 8 mm diameter holes are for rods along which can slide the width setting component.

Figure 5.19 shows three views of the depth-setting component. These are the views that would be seen in the directions of arrows X_2, Y_2 and Z_1 of Fig.5.18. These views have been chosen because they reveal more details than those in the directions of arrows X_1, Y_1 and Z_2. Students requiring practice in the use of drawing instruments should redraw these views, scale 2/1; the non-circular curves in the upper and left-hand views should be sketched in as neatly as possible. Alternatively, there is sufficient space around the views in Fig.5.19 for dimensions to be inserted.

Completely dimension the three-view drawing, scaling distances from Fig.5.19 when necessary. Be careful to take the function of the component into account and check that all necessary distances and angles have been included. Threads should be designated as shown in Fig.5.18. Add all necessary titles and labels.

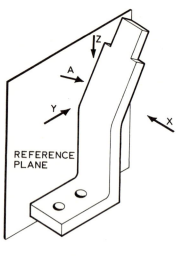

FIG. 6.1

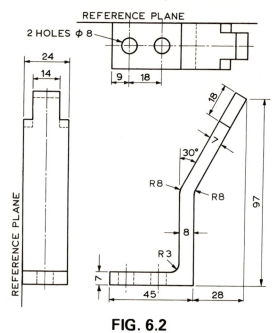

FIG. 6.2

THIRD ANGLE PROJECTION

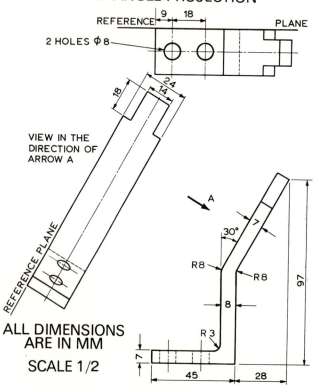

VIEW IN THE
DIRECTION OF
ARROW A

ALL DIMENSIONS
ARE IN MM

SCALE 1/2

FIG. 6.3

CHAPTER 6

Auxiliary Projections

Many engineering components have surfaces which are inclined to the main planes of the object. *Figure 6.1* is a pictorial view of such a component; the base (through which pass two holes) defines the three directions of any standard orthographic drawing. *Figure 6.2* is a drawing, in third angle projection, of the same component. The upper view is a view along arrow Z (of Fig.6.1), the lower right-hand view is a view along arrow X and the lower left-hand view is a view along arrow Y. None of the views shows the true shape of the inclined surface.

Notice that the dimensions which relate to the inclined surface have to be placed on adjacent views. The true 18 mm distance can only be shown on the lower right-hand view. It would be advantageous to have the three dimensions, which relate to the inclined surface, grouped together. This could be done if a view were drawn of what would be seen along arrow A of Fig.6.1, where arrow A is perpendicular to the inclined surface. Such a view is known as an AUXILIARY VIEW.

Figure 6.3 is an orthographic drawing showing views along arrows X, Z and A of Fig.6.1. The first thing to notice is that the rule relating to (in this case) third angle projection is still followed. The view along arrow X is the link view (see Chapter 1); and what would be seen from vertically above

it is drawn vertically above it. What would be seen looking from above view X and perpendicular to the sloping surface is drawn above view X and projected in a direction perpendicular to the sloping surface.

As the auxiliary view is rather special, it has an arrow to indicate the direction in which it is viewed and a title which refers to the arrow. The auxiliary view takes all the dimensions relating to the sloping surface and this makes the drawing easier to follow.

Now consider the construction of the auxiliary view. Fig.6.1 has a reference plane which has been indicated, as a line, in the appropriate views of Figs. 6.2 and 6.3. Looking first at Fig.6.2, it is clear that distances perpendicular to the reference plane will appear to have the same true length whether viewed along arrow Y or arrow Z. For example, the 24 mm width of the component is the same in views along Y and Z. In other words, rotating the viewing direction through 90°, about the X arrow, has no effect on distances parallel to X. The same is true if the viewing direction is moved from Y to A: distances parallel to X remain the same.

A convenient way to remember this rule is that DIMENSIONS, ALONG ALL SETS OF PROJECTION LINES RADIATING FROM A COMMON VIEW, ARE IDENTICAL. Thus, in Fig.6.2, there are two sets of imaginary projection lines, one horizontal

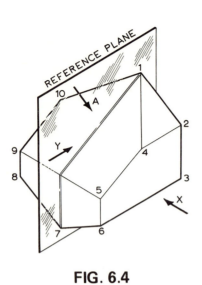

FIG. 6.4

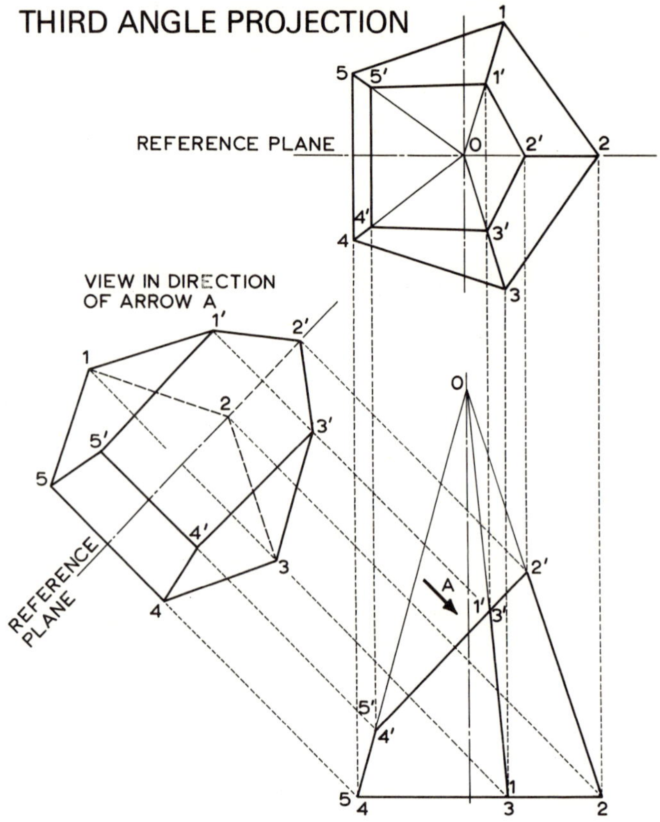

THIRD ANGLE PROJECTION

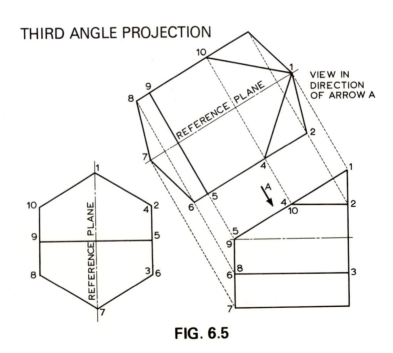

REFERENCE PLANE

VIEW IN DIRECTION OF ARROW A

REFERENCE PLANE

FIG. 6.6

THIRD ANGLE PROJECTION

VIEW IN DIRECTION OF ARROW A

REFERENCE PLANE

REFERENCE PLANE

FIG. 6.5

and the other vertical, radiating from the lower right-hand view. The reference plane can be drawn in at any convenient distance from the common view, and then dimensions, along both sets of projection lines perpendicular to the reference plane, must be identical.

The same is true of Fig.6.3: there are two sets of imaginary projection lines, one vertical and the other at 30° above the horizontal, radiating from the lowest view. Dimensions, along both sets of projection lines, are identical.

This important principle is further illustrated in *Figures 6.4* and *6.5*. A hexagonal prism is cut at an angle, as shown. A view along arrow X shows the angle at which the prism is cut. A view along arrow Y shows the cross-section of the prism. An auxiliary view along arrow A is needed to show the true shape of the inclined surface.

Imagine that the views along arrows X and Y are the only ones drawn in Fig.6.5; the auxiliary view is yet to be drawn. A perpendicular to the sloping surface specifies the direction for a new set of projection lines. The view along arrow X will then have two sets of projection lines radiating from it. In this example it is convenient to use a plane of symmetry as the reference plane. It can be drawn in, perpendicular to the dotted projection lines, at any convenient distance from the common view. The various points around the object are numbered for ease of identification and the perpendicular distance from the reference plane is transferred from the left-hand view to the auxiliary view. The appropriate lines are then drawn in and the surface 1-4-5-9-10 appears in its true shape in the auxiliary view. One other important fact is well illustrated in Fig.6.5; notice that the sense of the transferred distances remains the same. For example, in both the left-hand and the auxiliary view, points 8, 9 and 10 are the furthest from the common view, whereas points 2, 3, 4, 5 and 6 are the nearest

to the common view.

Figure 6.6 is a further example of the process of producing an auxiliary view that shows the true shape of a sloping surface. Basically, the object is a pyramid with a regular pentagonal base. It has been cut to produce the sloping surface shown in the lower view. The two right-hand views have been produced by constructing the complete pyramid and then locating the five points of the sloping surface by the intersection of the appropriate lines. For example, point 1' must lie on line 0-1 in both the upper and lower views. Point 1' is easily located when the sloping surface is drawn on the lower view and then the intersection of a vertical projection line up from 1' and the line 0-1, in the upper view, locates the position of 1' in the upper view.

Like Fig.6.5, the pyramid has a plane of symmetry and it is convenient to use it as the reference plane for the auxiliary view. A perpendicular to the sloping surface gives the direction for the projection lines to the auxiliary view. There are two sets of projection lines (indicated by dotted lines) from the lower view, and dimensions along these lines must be identical. Therefore the perpendicular distances from points 1, 1', 2, 2', 3, 3', 4, 4', 5, 5' to the reference plane in the upper view are transferred to the auxiliary view. Once again, the sense of the dimensions is not affected by the transfer; for example, point 3 is, in both cases, the nearest point to the common view.

EXERCISE 6.1

Figure 6.7 has two views of a cruciform section bar. With the aid of the 5 mm spaced guide lines, draw an auxiliary view as seen in the direction of arrow A, to show the true shape of the sloping surface. Do not show hidden detail. Label the auxiliary view. When you have finished, compare your solution with *Figure 6.8*.

THIRD ANGLE PROJECTION

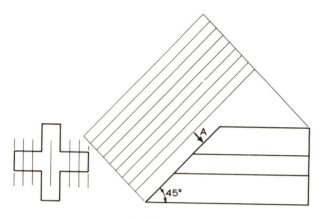

45°

A

FIG. 6.7

THIRD ANGLE PROJECTION

ALL DIMENSIONS
ARE IN MM

B

21

30°

15 10 15

15

5

5

10

5

15

5

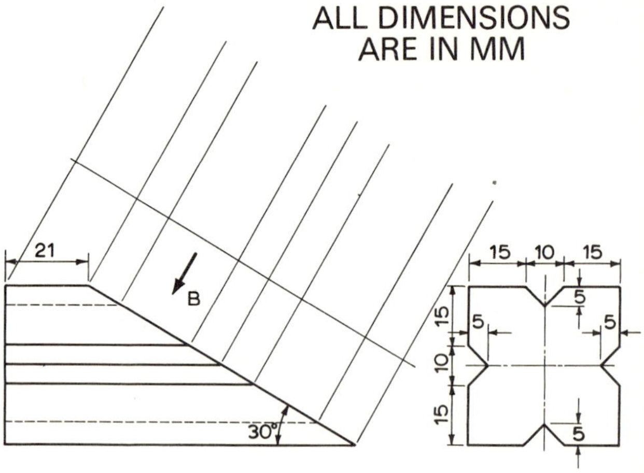

FIG. 6.9

THIRD ANGLE PROJECTION

VIEW IN DIRECTION
OF ARROW A

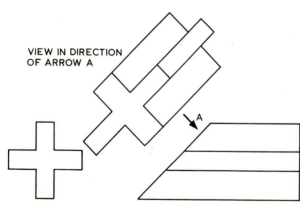

A

FIG. 6.8

THIRD ANGLE PROJECTION
ALL DIMENSIONS ARE IN MM

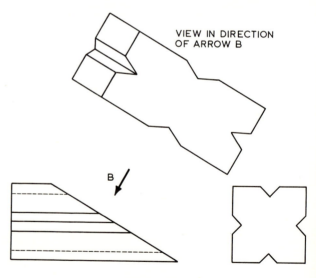

VIEW IN DIRECTION
OF ARROW B

B

FIG. 6.10

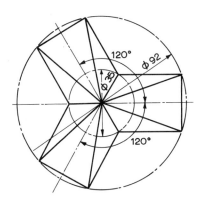

THIRD ANGLE PROJECTION
ALL DIMENSIONS ARE IN MM
SCALE 1/2

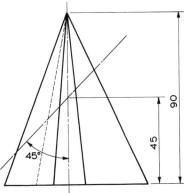

FIG. 6.11

THIRD ANGLE
PROJECTION

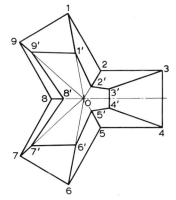

VIEW IN DIRECTION
OF ARROW C

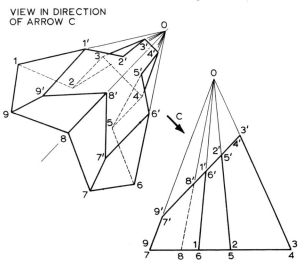

FIG. 6.12

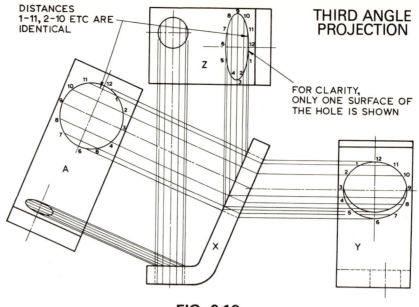

DISTANCES 1-11, 2-10 ETC ARE IDENTICAL

THIRD ANGLE PROJECTION

FOR CLARITY, ONLY ONE SURFACE OF THE HOLE IS SHOWN

FIG. 6.13

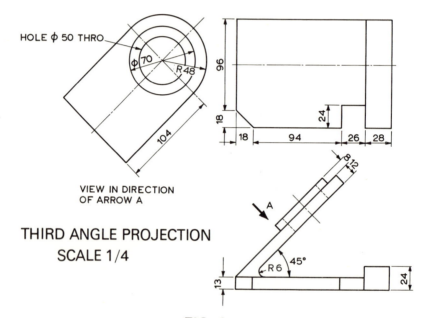

HOLE φ 50 THRO

VIEW IN DIRECTION OF ARROW A

THIRD ANGLE PROJECTION
SCALE 1/4

FIG. 6.14

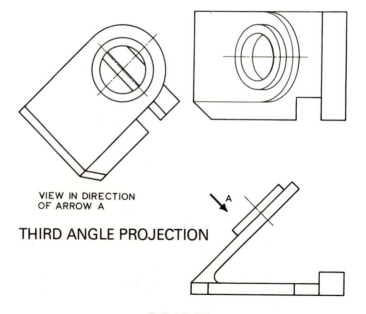

VIEW IN DIRECTION OF ARROW A

THIRD ANGLE PROJECTION

FIG. 6.15

EXERCISE 6.2

Figure 6.9 shows two views of a square section block with V grooves. Project an auxiliary view as seen in the direction of arrow B, to show the true shape of the sloping surface. No dimensions or hidden details are required. Label the auxiliary view. When you have finished, compare your solution with *Figure 6.10*.

EXERCISE 6.3

Figure 6.11 gives two views of a pyramid with a regular three-armed base. The pyramid is to have its top removed by cutting at the angle shown. Redraw the given views, scale 1/1, showing the effect of cutting the pyramid, and project an auxiliary view to show the true shape of the sloping surface. Show all hidden detail but do not add dimensions. Label the auxiliary view and state the type of projection used. When you have finished, compare your solution with *Figure 6.12*.

Circles

One of the disadvantages of auxiliary projection can be seen by referring back to Fig.6.3 (p.48). The two circular holes appear as ellipses in the auxiliary view and they have been constructed by a process that is rather time-consuming. When drawing an auxiliary view it is necessary to ask yourself: what purpose does the view serve? If, as in Fig.6.3, the important part of the view is the sloping surface, then only that part need be shown and the imaginary break can be indicated with an irregular thick line. If the holes have to be shown, as might be the case if only two views were to be drawn, a method of construction is shown in *Figure 6.13*. A different object has been used in order to illustrate a number of principles.

The view labelled X can be completed first as there are no unusual features. It is then possible to draw the straight lines of view Z and the small hole, but the large one cannot be drawn. In the view labelled A, it is possible to draw the large hole but not the small one; the circular outline of the large hole has been marked with twelve equally spaced, numbered points. A line from each point is projected back to the view X and another line is projected from view X to view Z. As always, distances across the object, measured along two lines radiating from the same point on view X, must be the same and so the 6-12 line may be located in view Z and the perpendicular distances from this line to the other ten points are transferred from view A to view Z. The symmetrical arrangement of the points reduces the amount of work to be done. The points are joined by using a curve template or sketching.

The same method is used to transfer distances from the small circle in view Z to construct the ellipse in view A. Normally the projection lines should be erased.

The view labelled Y has been added to illustrate the importance of projecting the ellipses from the correct planes. In this view the complete ellipse comes from the right-hand surface of view X and the partial ellipse comes from the left-hand surface. Notice also that the numbered points now appear anticlockwise. The symmetry of the ellipse disguises this inversion but it is bound to arise if the ellipse is properly constructed by maintaining the sense of the dimensions (for example, in views A, Z and Y, point 3 is always the one nearest to view X).

EXERCISE 6.4

Figure 6.14 shows a three-view drawing of a bracket; two of the views are incomplete. Redraw the views, scale 1/1, and complete both the auxiliary view and the upper right-hand view. Do not add dimensions or hidden detail. Show all necessary titles and labels. When you have finished, compare your solution with *Figure 6.15*.

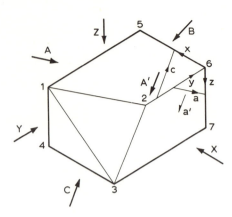

FIG. 6.16

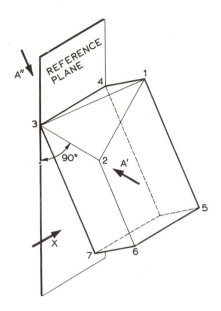

FIG. 6.18

THIRD ANGLE PROJECTION

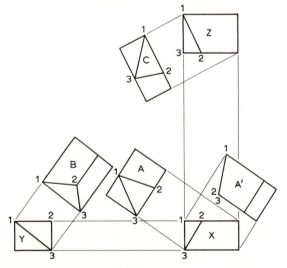

FIG. 6.17

THIRD ANGLE PROJECTION

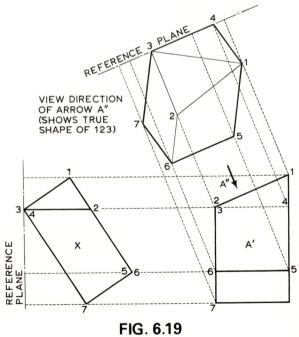

VIEW DIRECTION
OF ARROW A"
(SHOWS TRUE
SHAPE OF 123)

FIG. 6.19

Double auxiliary projection

The sloping surfaces examined in this chapter so far have been inclined to only one of the projection planes. In other words, there has always been one view in which the sloping surface appeared as a straight line. *Figure 6.16* shows an object with a triangular plane (1-2-3) which is inclined to two projection planes. The projection planes of Fig.6.16 are those normal to *x*, *y* and *z*, defined by directions 6-5, 2-6 and 6-7 respectively. The plane 1-2-3 will not appear as a straight line when viewed in any of these directions.

Figure 6.17 demonstrates this point. Look first at views X, Y and Z; the plane appears as a different shape of triangle in each of them. Furthermore, it is impossible to draw an auxiliary view which will show the true shape of the triangle. An auxiliary view has been projected from view X in a direction perpendicular to line 2-3 (direction "a" of Fig.6.16); it shows the true length of line 2-3 but the other lines are foreshortened. An auxiliary view has been projected from view Y in the direction perpendicular to line 1-3: it only shows the true length of line 1-3. Similarly, the view in the direction of arrow C, projected from view Z, shows only the true length of line 1-2.

Clearly, it is not possible to produce, directly, an auxiliary view showing the true shape of the triangular surface. What is missing is a view in which the triangular surface appears as a straight line. Such a view may be obtained by an auxiliary projection <u>along</u> rather than perpendicular to one of the lines of the triangular surface. For example, a view along arrow A' (of Fig.6.16) will make points 2 and 3 appear coincident and superimpose lines 1-2 and 1-3. This auxiliary view has to be projected from view X since it is the view where line 2-3 appears in its true length. Thus, view A' provides the first stage in the production of a view of the true shape of the triangular surface.

Figure 6.18 illustrates the same object, differently orientated, and with a reference plane shown which is perpendicular to line 2-3. Views X and A' from Fig.6.17, have been enlarged and redrawn in *Figure 6.19*. Try to think of these views as the original views of the object and forget the other views of Fig.6.17. The lower views of Fig.6.19 contain all the information needed for the required view.

Referring now to Fig.6.18, it is possible to see, whether one looks along arrow X or arrow A" because they are both parallel to the reference plane, that distances perpendicular to the reference plane must remain the same. The auxiliary view along arrow A" is drawn by the usual method of projecting in a direction perpendicular to line 1-2, drawing the reference line in a convenient position and parallel to line 1-2, and transferring dimensions taken along the projection lines on view X to the auxiliary view. This view, along arrow A", shows the true shape of the triangular surface.

The process of drawing one auxiliary view and then projecting another auxiliary view from it is known as DOUBLE or SECOND AUXILIARY PROJECTION. Apart from finding the true shape of a surface not appearing as a line in any of the original views, the process has a number of other uses and some of them will be examined below.

It may be shown that there is no theoretical limit to the number of auxiliary views that could be drawn. There are problems where Treble Auxiliary Projection is a convenient method of analysis but, for all practical purposes, there is no need to use anything more than a Double Auxiliary Projection.

Now examine the process applied to an engineering component rather than a geometrical block. In the lower right-hand corner of *Figure 6.20* are two views of a simple bracket (for clarity, any holes have been omitted). Neither view shows the true shape of the inclined surface and the first step towards producing such a view is to draw an auxiliary view that shows the surface as a straight line.

This can be done by selecting a viewing direction along any line in the

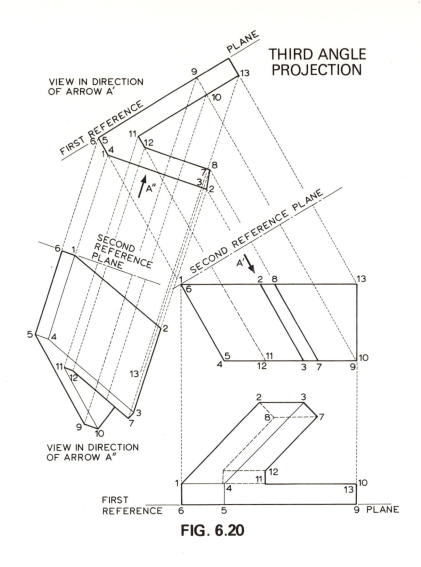

THIRD ANGLE
PROJECTION

VIEW IN DIRECTION
OF ARROW A'

FIRST REFERENCE

PLANE

SECOND
REFERENCE
PLANE

SECOND REFERENCE PLANE

VIEW IN DIRECTION
OF ARROW A"

FIRST
REFERENCE

PLANE

FIG. 6.20

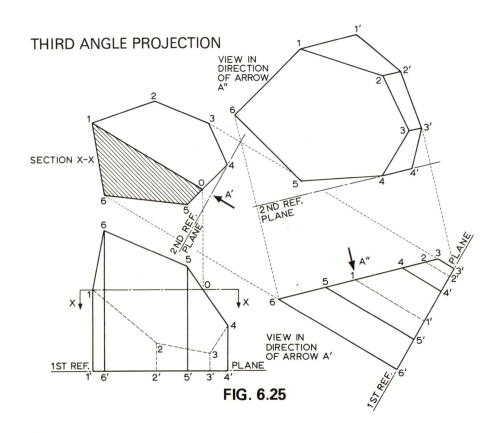

THIRD ANGLE PROJECTION

VIEW IN
DIRECTION
OF ARROW
A"

SECTION X-X

2ND REF.
PLANE

2ND REF. PLANE

PLANE

VIEW IN
DIRECTION
OF ARROW A'

1ST REF.   PLANE

1ST REF.

FIG. 6.25

surface which appears in its true
length. In this case, lines 1-4 and
2-3 in the lower view, or lines 2-3 and
6-4 in the upper view, are equally suitable but the upper view has been used
in order to fit all the views into the
available space.

A view along arrow A' is drawn in
the usual way and dimensions perpendicular to the First Reference Plane in
the lower view are transferred to the
appropriate projection in the First
Auxiliary View. The surface 1-2-3-4
appears as a straight line.

A second auxiliary view is projected, perpendicular to plane 1-2-3-4.
The first auxiliary view is now the
common view from which two sets of projection lines radiate; two new reference planes are introduced, perpendicular to the two sets of lines. Dimensions are transferred from the upper
view to the second auxiliary view. In
this example, only visible lines have
been shown but it is necessary to locate point 13 so that the 10-13 line is
in the correct direction.

The great difficulty with this
type of problem is visualization. The
principles are extremely simple; it is
the odd angles and orientations of the
object that create challenges for even
the most experienced draughtsman.
There are three points to remember:
first, the basic rule, DIMENSIONS,
ALONG ALL SETS OF PROJECTION LINES
RADIATING FROM A COMMON VIEW, ARE
IDENTICAL. Second, points on views
projected from a common view always re-
tain their relative distances from the
common view. Third, it is very helpful to number the important points
around the object, for ease of identification. If you keep these things in
mind, you should be able to solve most
double auxiliary projection problems,
after some practice.

EXERCISE 6.5

Figure 6.21 is a pictorial view
of a component. Draw, scale 2/1, in
third angle projection, views as seen
in the direction of arrows Y and Z.
(N.B. Lines 1-2, 3-4, 5-6, 7-8 and
9-10 are all parallel.) Hence, draw
an auxiliary view in which surface
5-6-7-8 appears as a straight line,
and a second auxiliary view showing
the true shape of this surface. Do not
add hidden details or dimensions. Insert all appropriate titles and labels.
When you have finished, compare your
solution with *Figure 6.23*.

EXERCISE 6.6

Figure 6.22 is a pictorial view of
a bracket. Draw, scale 2/1, in third
angle projection, views as seen in the
direction of arrows X and Y. (N.B.

Points P, Q, R and S are all 40 mm
above the base plane.) Hence draw an
auxiliary view in which the L-shaped
surface appears as a straight line, and
a second auxiliary view showing the
true shape of this surface. Do not add
dimensions and show only the hidden
details required for the construction
of the auxiliary views. Add all appropriate titles and labels. When you
have finished, compare your solution
with *Figure 6.24*.

Surfaces with all sides foreshortened

In the above examples of double
auxiliary projection, at least one side
of the sloping surface has appeared in
its true length. Such a side can be
used to determine the direction for a
first auxiliary view which will show
the surface as a straight line. A more
difficult problem arises when none of
the sides appears in its true length
and the first task is to find the direction in which to project the first auxiliary view.

Figure 6.25 contains such a surface, 1-2-3-4-5-6, and shows no side
in its true length in either of the
two left-hand views. Consider a horizontal section through point 1 - Section
X-X. By projecting vertically upwards
from the point O, where this section
plane cuts the line 4-5, the section
X-X can be drawn; line O-1 must be a
horizontal line on the sectional view
which passes through two points on the
sloping surface. An auxiliary view,
in the direction of arrow A', along O-1,
will make the whole surface, 1-2-3-4-5-6,
appear as a straight line.

This auxiliary view has been drawn
in the usual way, and a second auxiliary view can then be drawn as seen in
the direction of arrow A", perpendicular
to the line 6-5-1-4-2-3. The view in
the direction of arrow A" shows the true
shape of the sloping surface.

The most important new principle
here is the method of finding the direction in which to project the first auxiliary view. As in the previous examples, a line in the surface of the plane
is required which appears in its true
length. The upper left-hand view of
Fig.6.25 will show only horizontal
lines in their true length and so it
is necessary to create a horizontal
line in the lower view and to project
it back to the upper view.

EXERCISE 6.7

Figure 6.26 is a pictorial view of
a component with a special surface,
1-2-3-4. Using double auxiliary projection, construct the true shape of the
special surface.

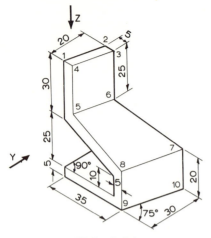

ALL DIMENSIONS
ARE IN MM

FIG. 6.21

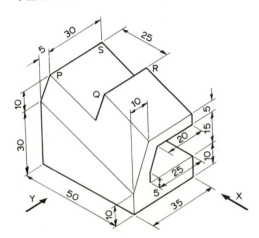

ALL DIMENSIONS ARE IN MM

FIG. 6.22

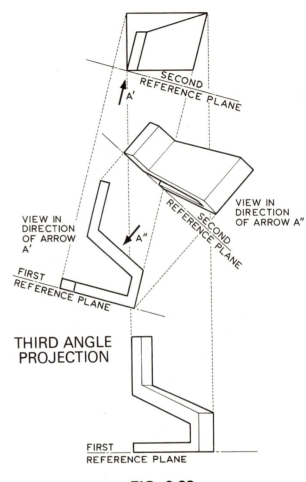

THIRD ANGLE
PROJECTION

FIG. 6.23

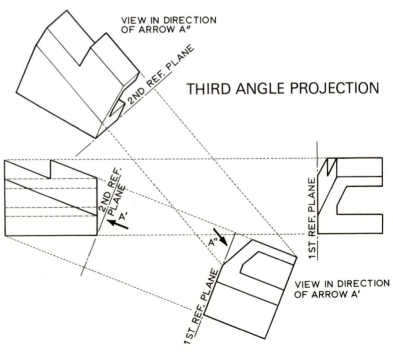

THIRD ANGLE PROJECTION

FIG. 6.24

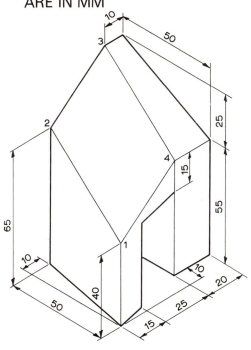

ALL DIMENSIONS
ARE IN MM

FIG. 6.26

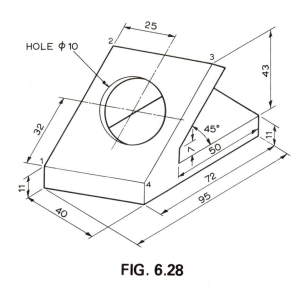

ALL DIMENSIONS ARE IN MM

FIG. 6.28

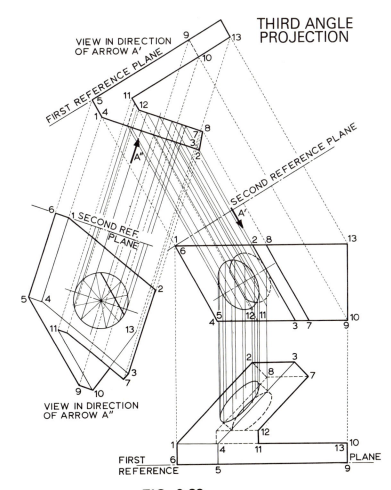

FIG. 6.29

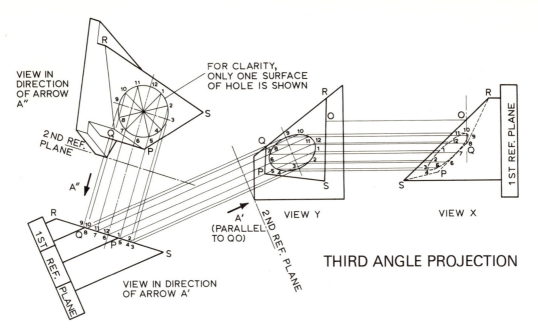

FIG. 6.27

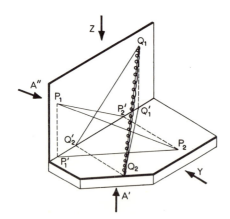

FIG. 6.30

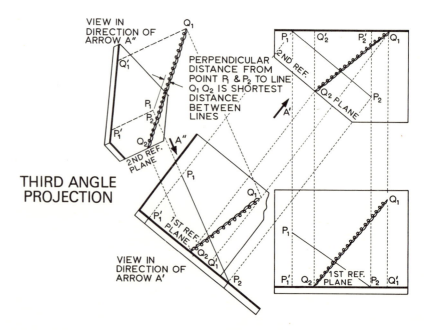

FIG. 6.31

Circles in double auxiliary projection

In principle, there is no additional difficulty with circles when more than one successive auxiliary views are involved. *Figure 6.27* shows, on the right-hand side, two views of an irregular block with a trapezium-shaped base; these views will be considered to be the original views and have been labelled VIEW X and VIEW Y.

The drawing is to show a circular hole, drilled with its axis perpendicular to the inclined quadrilateral surface. Like Fig.6.25, the first step is to discover the direction of an auxiliary view that will show the sloping surface as a straight line. This has been done by drawing a vertical line through Q (in VIEW X), to cut line SR at O; O is then projected horizontally to VIEW Y, in which QO gives the direction for a first auxiliary view.

The first auxiliary view is drawn in the usual way and then a second auxiliary view is projected from it in a direction perpendicular to the line RQPS. In the view in the direction of arrow A", the surface PQRS appears in its true shape and it is now possible to draw a circle to represent the hole in the required position.

The circle is marked with twelve equally spaced points and the line 9-3 is parallel to the second reference plane. The twelve points are projected back to the view in the direction of arrow A', to meet the line RQPS. For clarity in this example, the depth of the hole will not be considered.

The next step is to draw the ellipse in VIEW Y, and this is done in just the same way as in Fig.6.12. The dimensions from the "2nd Ref. Plane" to line 9-3, and the perpendicular distances of the other ten points from line 9-3, are identical in both the views projected from the view in the direction of arrow A'.

Finally, the hole is shown as a hidden shape in VIEW X. Each of the twelve points is the same perpendicular distance from the "1st Ref. Plane" in both the views projected from VIEW Y.

It can be seen that it is more complicated to draw circles in double auxiliary projection, but no new principles are involved.

EXERCISE 6.8

Figure 6.28 shows the same basic object as Fig.6.20; a hole has been added and its axis is perpendicular to the surface 1-2-3-4. Redraw the four views of Fig.6.20 showing the visible and hidden edges of the hole in each view. Add all necessary titles and labels but do not insert dimensions. When you have finished, compare your solution with *Figure 6.29*.

Projective geometry

So far in this Chapter, double auxiliary projection has been used as a method of finding the true shape of plane surfaces: this is not only a useful process in engineering drawings, but can also be used for analysing the geometry of proposed designs. For example, the true shape of a surface gives all the correct distances <u>and</u> <u>angles</u> on that surface. A view <u>in</u> which two intersecting surfaces both appear as straight lines gives the angle between the surfaces.

The subject is so vast that only a separate book could do justice to it. However, the following example is a particularly useful application of double auxiliary projection to an essentially geometrical problem.

Figure 6.30. shows two perpendicular boards; the surface of the thin one is a vertical plane and the surface of the thick one is a horizontal plane. Two straight lines, P_1P_2 and Q_1Q_2, are defined by the positions of P_1, Q_1 and P_2, Q_2 in the vertical and horizontal planes, respectively.

The line Q_1Q_2 is shown with a coil around it for ease of identification. The four points have been projected horizontally or vertically on to planes perpendicular to those in which they lie.

The two lines do not intersect and it is required to find the shortest distance between them. Such a problem frequently arises when power lines and other cables are in close proximity, or when the locus of a moving object brings part of it close to a stationary component.

Figure 6.31 shows, on the right-hand side, views of the two boards and lines, as seen in the direction of arrows Y and Z, in third angle projection. Notice that the horizontal board has a corner removed to produce a vertical plane, parallel to the plane $P_1P_1'P_2$.

The shortest distance between P_1P_2 and Q_1Q_2 will appear when either line is viewed in a direction that makes its two points coincide. Arrow A" (in Fig. 6.30) is such a direction, and the first step in constructing such a view is to draw an auxiliary view in which line P_1P_2 appears in its true length.

Arrow A' is perpendicular to the plane $P_1P_1'P_2$ and an auxiliary view in the direction of this arrow has been drawn in the usual way. The surface of the horizontal board is used as the "1st Ref. Plane" and dimensions perpendicular to this plane are transferred from the view in the direction of arrow Y to the view in the direction of arrow

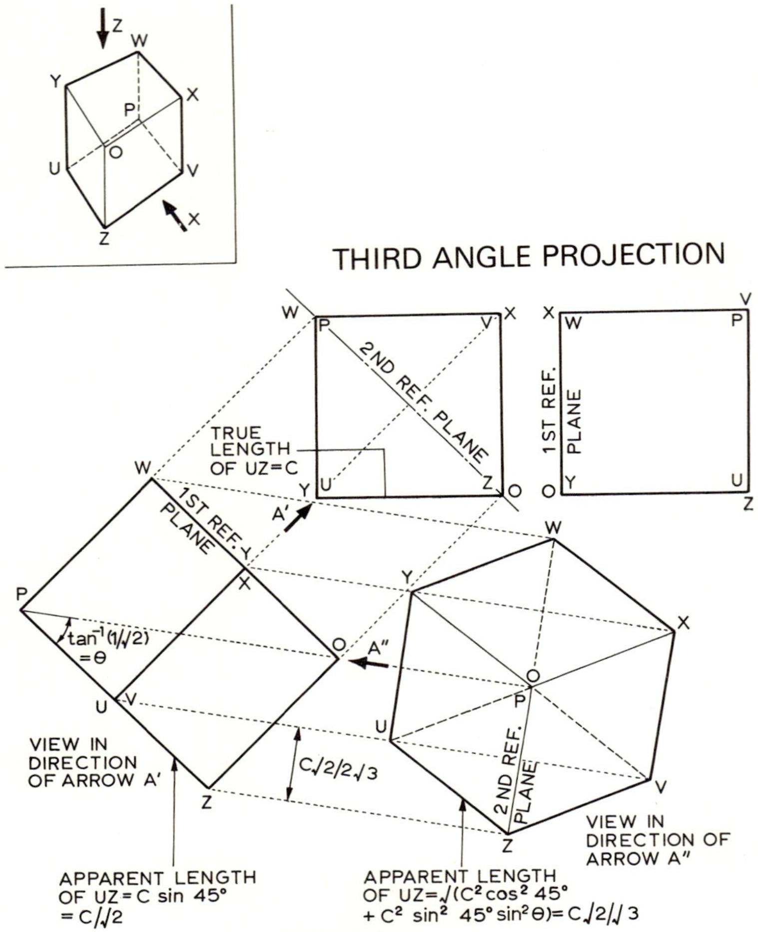

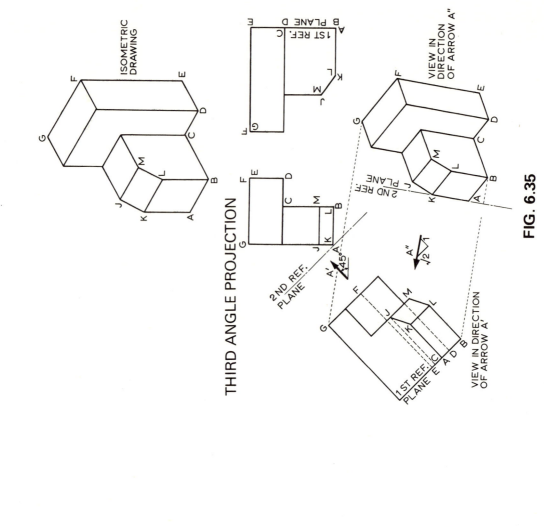

THIRD ANGLE PROJECTION

ISOMETRIC DRAWING

1ST REF. PLANE D

VIEW IN DIRECTION OF ARROW A"

2ND REF. PLANE

VIEW IN DIRECTION OF ARROW A'

2ND REF. PLANE

1ST REF. PLANE E

FIG. 6.35

THIRD ANGLE PROJECTION

2ND REF. PLANE

1ST REF. PLANE

VIEW IN DIRECTION OF ARROW A"

VIEW IN DIRECTION OF ARROW A'

TRUE LENGTH OF UZ=C

APPARENT LENGTH OF UZ$=\sqrt{(C^2 \cos^2 45°} + C^2 \sin^2 45° \sin^2 \theta)=C\sqrt{2}/\sqrt{3}$

APPARENT LENGTH OF UZ=C sin 45° $=C/\sqrt{2}$

$\tan^{-1}(1/\sqrt{2})=\theta$

$C\sqrt{2}/\sqrt{3}$

FIG. 6.34

A', in which line P_1P_2 appears in its true length.

The vertical board will now be ignored as its existence would hide P_1 and P_2 from the viewer who looked in the direction of arrow A". The second auxiliary view is drawn by using the vertical plane through point Q_2 as the "2nd Ref. Plane" and dimensions perpendicular to this plane are transferred from the view in the direction of arrow Z to the view in the direction of arrow A". Line P_1P_2 now appears as a point and the shortest distance between the two lines is the perpendicular distance from that point to line Q_1Q_2.

If required, the point on Q_1Q_2, closest to P_1P_2, can be located and projected back to the view in the direction of arrow A', in which view the corresponding point on P_1P_2 can be located. Both points can then be projected back to the two original views on the right-hand side.

EXERCISE 6.9

Figure 6.32 shows a cable, P_1P_2, passing close to the edge of a roof, Q_1Q_2. Use double auxiliary projection to determine the shortest distance between P_1P_2 and Q_1Q_2. Locate, in all views, the points on P_1P_2 and Q_1Q_2 which are closest together. Compare your solution with *Figure 6.33*.

Isometric projection

The method for producing pictorial drawings, known as "Isometric Drawing", is examined in Chapter 8. It is convenient to discuss the origin of the term "Isometric" in this chapter, as it involves a process of double auxiliary projection.

Figure 6.34 is a drawing of a cube; a pictorial view of the cube is inset at the top left and its corners are lettered for ease of identification. In addition, two of the three visible planes are shaded in a distinctive way.

At the top of the orthographic drawing are views as seen in the directions of arrows Z and X. The auxiliary views have been drawn so as to produce a view of the cube as seen in a direction from

O to P. The first auxiliary view has been drawn to show the diagonal PO in its true length and the second auxiliary view is projected along PO and gives the desired result.

It may be shown that all the visible and hidden square surfaces of the cube become congruent rhombuses in the view in the direction of arrow A". The acute angles of the rhombuses are 60°. The edges of the cube appear shorter than their true length by a factor of $\sqrt{2}/\sqrt{3}$.

Hence, an "Isometric" (meaning "Equal-length") projection is a view like one along the long diagonal of a cube, when equal lengths in three mutually perpendicular directions, OX, OY and OZ, have equal apparent lengths in the view that is constructed. As the effect of an isometric projection is now known, a pictorial view of any object can be produced without using double auxiliary projection. The isometric projection can be produced directly by reducing distances in the directions OX, OY and OZ by a factor of $\sqrt{2}/\sqrt{3}$ and setting these distances out along axes with angles of 60° between them. Indeed, if a picture of the object is all that is required, the factor $\sqrt{2}/\sqrt{3}$ can be ignored and the only effect will be that the object will appear rather larger than life.

Figure 6.35 illustrates the direct construction of an isometric drawing and the construction of the same view (with, of necessity, a different orientation) by the use of a double auxiliary projection. In the isometric drawing, the dimensions in the three directions OX, OY and OZ have been transferred, without any reduction, from the two upper views of the orthographic drawing. The four views in the orthographic drawing are arranged in exactly the same way as the views of Fig.6.34. Follow through the successive auxiliary views and make sure that you understand how they have been constructed.

Isometric drawing is treated in detail in Chapter 8.

EXERCISE 6.10

Make an isometric projection from the two views of the object shown in Fig.6.8.

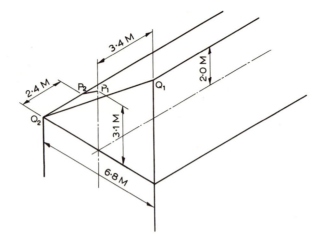

FIG. 6.32

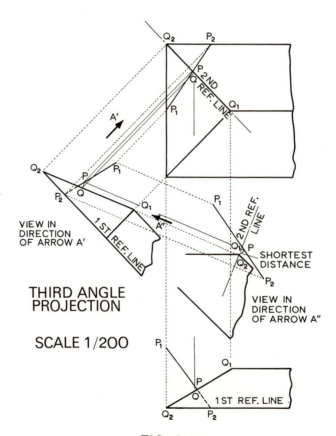

FIG. 6.33

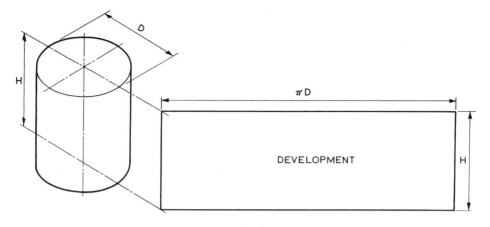

FIG. 7.1

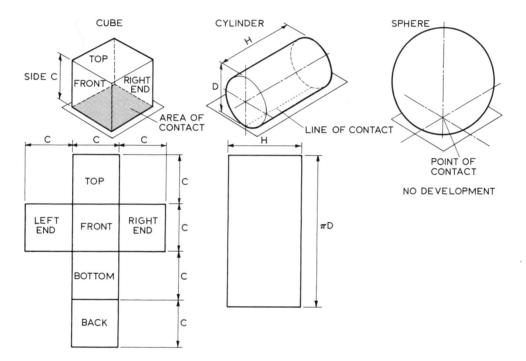

FIG. 7.2

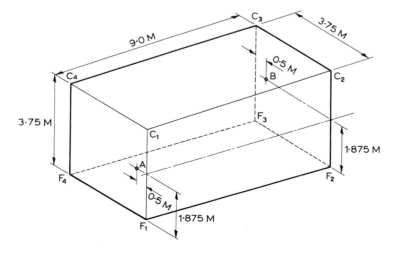

FIG. 7.3

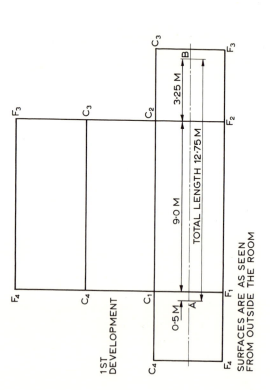

FIG. 7.4

1ST DEVELOPMENT

SURFACES ARE AS SEEN FROM OUTSIDE THE ROOM

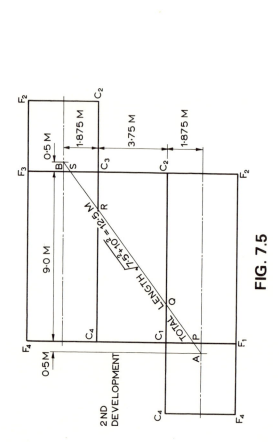

FIG. 7.5

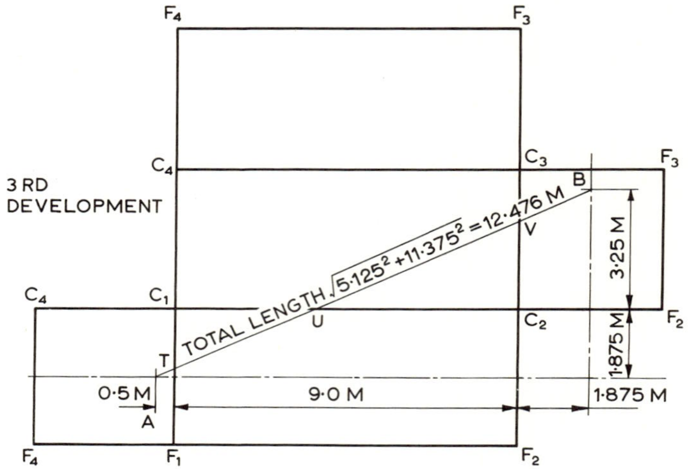

2ND DEVELOPMENT

FIG. 7.6

3RD DEVELOPMENT

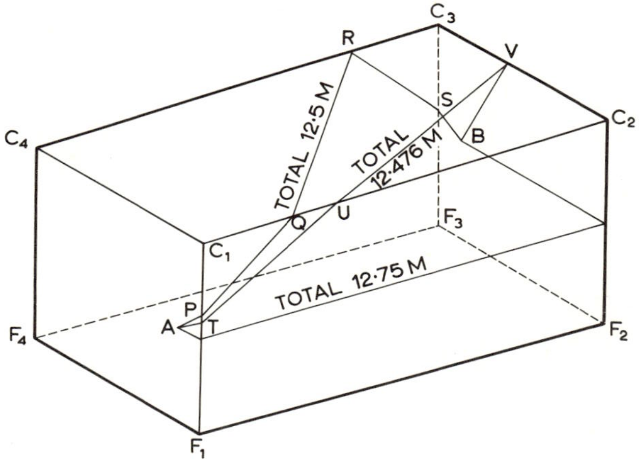

FIG. 7.7

CHAPTER 7

Developments

The development of a three-dimensional surface is the two-dimensional shape which can be converted into that surface. For example, *Figure 7.1* shows a cylindrical surface of height H and diameter D. Clearly, the development of this surface is the rectangle of height H and length πD; a paper rectangle of this size could be converted into the cylindrical surface.

There are two important points to notice about the surface and its development. Firstly, if the solid surface were placed on a plane surface, there would be line contact between the two surfaces. Secondly, the shortest distance between any two points, measured on the solid surface, is the same on the development.

Figure 7.2 illustrates three types of solid surface. The cube consists entirely of flat surfaces and has no curvature; it can be developed and one possible development is shown. The cylinder has line contact with a plane surface; it can be developed and this is shown. The cylinder has <u>single</u> curvature, the test for which is the line contact with a plane surface. The sphere has point contact with a plane surface and this is the test for <u>double</u> curvature; it cannot be developed.

There are two practical uses for the idea of development of surfaces. In a manufacturing process, the development may be cut out of a sheet of material and then the material is bent and joined to produce the solid surface. In a design situation, it may be necessary to examine the geometry of points and lines on a solid surface. If the surface can be developed, then the design problem is easier to handle and to understand. Navigation is a difficult subject to most people because the earth is very nearly spherical and a spherical surface cannot be developed. The type of projections used for maps cannot produce a development of the surface, and distances between any two points do not always correspond with the true distances. How much easier things would be if we lived on the surface of a cylinder!

Figure 7.3 shows an amusing idealised problem with some practical implications. The surfaces $F_1F_2F_3F_4$ and $C_1C_2C_3C_4$ represent the floor and ceiling of a room, respectively. What is the shortest length of cable that could be used to connect point A with point B? The cable must remain in contact with a wall or the ceiling along its entire length.

At first sight, it seems that the shortest distance between points A and B is 12·75 m, the cable following a horizontal path along three walls. Such a path would be represented on a development of the room's surfaces by a horizontal line on *Figure 7.4*. However, this is only one of many ways of developing the room (the floor surface has been omitted as it is not the best place to run a cable and, in fact, is not needed for a solution) and you are asked to look at alternative developments to see if the length of cable can be reduced.

A second development is shown in *Figure 7.5*. If the cable followed the route shown, the length required would be exactly 12·5 m. Notice that the line AB crosses none of the discontinuities in the development and so the distance is the true length of a cable that crosses the edges at points P, Q, R and S.

A third development is shown in *Figure 7.6*; the length of cable required is slightly less than the last case, 12·476 m. Once again the line AB crosses no discontinuities and the distance is the true length of a cable that crosses the edges at points T, U and V. The three routes are shown pictorially in *Figure 7.7*. Distances such as C_1Q, C_3V, etc., could be scaled from the developments.

The above problem demonstrates that information can be more easily obtained from the development of a surface than from the surface itself, provided the various arrangements of the development are fully examined. A solid of this rectangular form is easily developed but the developments of the following examples are progressively more complex.

69

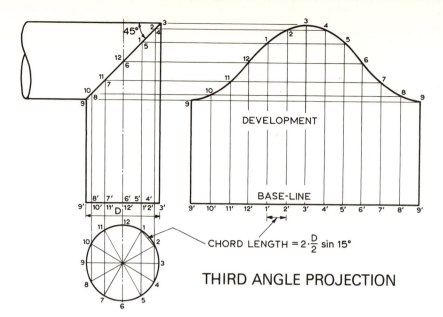

DEVELOPMENT

BASE-LINE

CHORD LENGTH = $2 \cdot \dfrac{D}{2} \sin 15°$

THIRD ANGLE PROJECTION

FIG. 7.8

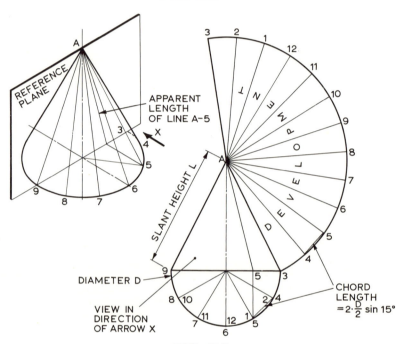

REFERENCE PLANE

APPARENT LENGTH OF LINE A-5

SLANT HEIGHT L

DEVELOPMENT

DIAMETER D

VIEW IN DIRECTION OF ARROW X

CHORD LENGTH = $2 \cdot \dfrac{D}{2} \sin 15°$

FIG. 7.9

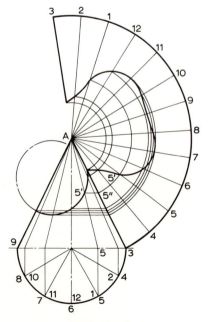

FIG. 7.10

Modified cylinder

Figure 7.8 shows the type of joint frequently used for two tubes which form a right-angled connection. The surface of the vertical tube has been developed by the following method:

(1) Divide the circular end view of the tube into 12 equally spaced points. Label the points 1 to 12.

(2) Project vertically upwards from each of the twelve points, across the upper view, to meet the interface of the two tubes.

(3) Set out a base-line for the development by making distances 9'-10', 10'-11', etc., equal to chords 9-10, 10-11, etc. (This results in a small error as the base of the development will be $12D.\sin15^{\circ} = 3 \cdot 106D$ in length, instead of πD, a reduction of about 1%.)

(4) Project upwards from each point on the base-line and across from the corresponding point on the interface. Join the points so found with a smooth curve.

This method produces a reasonably accurate development of the surface. Distances parallel with the axis of the tube are true distances which stay the same in the development. Any part of a cylinder can be developed by this method, provided it is possible to draw a view showing distances parallel to the axis. The spacing along the base-line need not be uniform, but chords subtending an angle of more than 30° at the centre of the circle are not recommended.

Right cone

Figure 7.9 contains a pictorial view of a right cone (i.e. the apex is perpendicularly above the centre of the circular base). A view as seen in the direction of arrow X is the isosceles triangle A-3-9, and this is also the triangle produced by the intersection of the cone and the reference plane, in the pictorial view.

Notice that only the generators A-3 and A-9 appear in their true length. The generator A-5 is foreshortened, but its true length can be obtained by projecting horizontally across to points 3 or 9, as all generators are of length L.

The surface of the cone has been developed by drawing a circular arc, centre A, radius L. Only half of the base is shown, using the line 3-9 as an axis of symmetry. The straight line distance between two points, such as 4 and 5, is transferred from the view of half the base to the development. The error introduced depends on the proportions of the cone but is rarely of any significance.

In effect, the cone has been divided into twelve small triangles with a common apex. The sides of the triangles are generators of the cone. The development is shown as a sector because <u>any</u> point on the base is distance L from the apex.

Figure 7.10 shows a right cone cut by a cylinder, the axis of which is parallel to arrow X of Fig.7.9. The bottom part of the cone has been developed by the following method:

(1) Draw a development of the complete cone (by the method illustrated in Fig.7.9).

(2) Where the cylinder intersects a generator, e.g. point 5' on line A-5, project horizontally to meet line A-3 at 5". Just as A-3 is the true length of line A-5, so A-5" is the true length of line A-5'.

(3) With centre A, radius A-5", draw an arc to cut the A-5 line of the development. (This is also taken to the A-1 line as points 5 and 1 are arranged symmetrically.)

(4) Repeat the process for the intersection of the cylinder and the other generators. Join the points with a smooth, symmetrical curve.

EXERCISE 7.1

Figure 7.11 is a pictorial view of a T-connection between two tubes of different diameters. The transition piece is part of a right cone. Draw, in orthographic projection, views as seen in the directions of arrows X and Y. Choose any suitable scale and make the tubes, away from the transition piece, of any convenient length.

Develop the surface of the cone and, by considering lines such as O_1-P-Q-R-S-O_2 in both views, develop the hole in the 52 mm diameter tube. When you have finished, compare your solution with *Figure 7.12*.

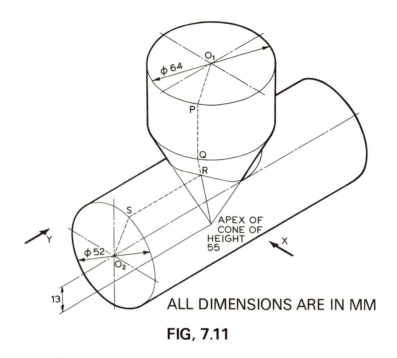

$\phi 64$

O_1

P

Q

R

S

$\phi 52$

O_2

13

APEX OF
CONE OF
HEIGHT
55

Y

X

ALL DIMENSIONS ARE IN MM

FIG, 7.11

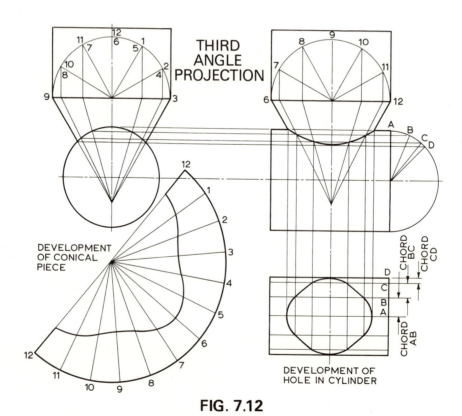

THIRD
ANGLE
PROJECTION

DEVELOPMENT
OF CONICAL
PIECE

CHORD BC

CHORD CD

CHORD AB

DEVELOPMENT OF
HOLE IN CYLINDER

FIG. 7.12

ALL DIMENSIONS ARE IN MM

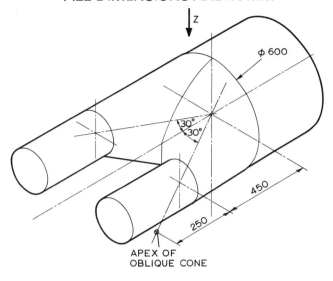

FIG. 7.15

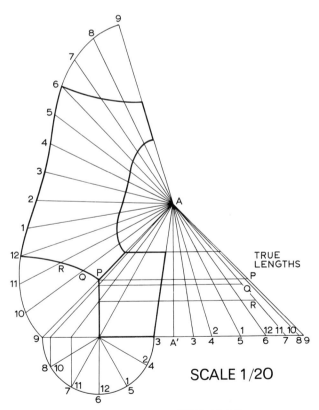

SCALE 1/20

FIG. 7.16

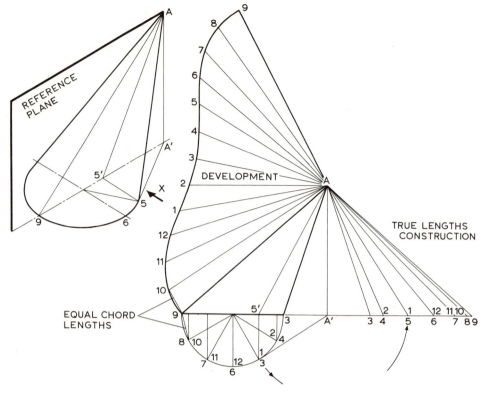

FIG. 7.13

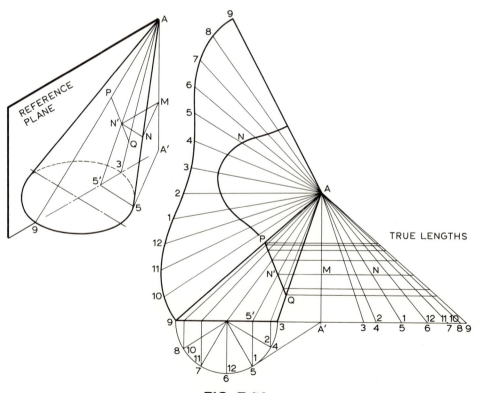

FIG. 7.14

Oblique cone

Figure 7.13 contains a pictorial view of an oblique cone (i.e. the apex is offset from a perpendicular through the centre of the circular base). A view as seen in the direction of arrow X is the triangle A-3-9, and this is also produced by the intersection of the cone and the vertical reference plane in the pictorial view.

Notice that only a line such as A-9, in the reference plane, appears in its true length. The line A-5 is foreshortened, its apparent length appearing as A-5'. The true length may be obtained by constructing the right-angled triangle A-A'-5; a convenient way of doing this is shown in the figure.

Only half of the base is shown, using the line 3-9 as an axis of symmetry. The apex A is projected on to the plane containing the base, to give point A'. With centre A', radius A'-5, the distance A'-5 is set out at 90° to the right of AA': the hypoteneuse of triangle A-A'-5 is the true length of line A-5. The same construction process is repeated, to draw a complete set of true length lines.

The surface of the cone has been developed by starting at line A-9, which already appears in its true length, and drawing the triangle A-9-10. Line A-10 comes from the set of true lengths and the side 9-10 equals the chord 9-10, from the circular base. Triangle A-10-11 comes next and the other triangles follow, in the appropriate order. The numbered points are joined with a smooth curve.

Notice that the position of the discontinuity in the development is of some importance when the manufacture of such a cone is being considered. The number of developed shapes that can be cut from a given size of sheet material would be affected by the position of the join. For example, a join along line A-3, instead of A-9, might enable more shapes to be cut from the same size of sheet.

Figure 7.14 shows an oblique cone, cut by a plane surface PQ; the intersection of this plane and the reference plane (in the pictorial view) produces the line P-Q. The lower part of the cone has been developed by the following method:

(1) Draw a development of the complete cone (by the method illustrated in Fig.7.13).
(2) Where the plane PQ intersects a line such as A-5' (point N'), project horizontally to meet the true length line A-5 at point N. Just as A-5 is the true length of line A-5', so A-N is the true length of line A-N'.
(3) With centre A, radius A-N, swing an arc to cut the A-5 line of the development. (This arc is also taken to the A-1 line, as points 5 and 1 are arranged symmetrically.)
(4) Repeat the process for the intersection of the plane PQ and the other lines on the surface of the cone. Join the points with a smooth, symmetrical curve.

EXERCISE 7.2

Figure 7.15 is a pictorial view of the most common application of oblique cones in piping. Two small diameter pipes are joined to form a single, large diameter, pipe; the axes of all three pipes are parallel. The transition piece takes the form of two intersecting oblique cones with a common base. When viewed in the direction of arrow Z, one half of the transition piece is the mirror image of the other.

Draw a view of one oblique cone, as seen in the direction of arrow Z, and hence develop the surface of one half of the transition piece. When you have finished, compare your solution with *Figure 7.16*. (It may be helpful if the development in Fig.7.14 is traced, cut out and joined up to give a truncated oblique cone, which may be used for an aid in this exercise. Indeed, any of the solutions, in this chapter, may be used in this way if difficulties are encountered.)

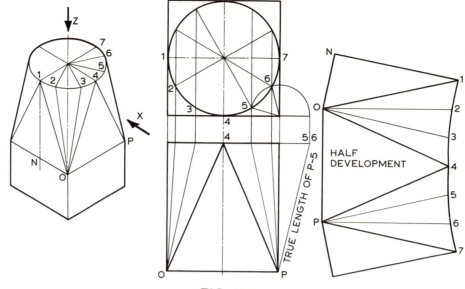

FIG, 7.17

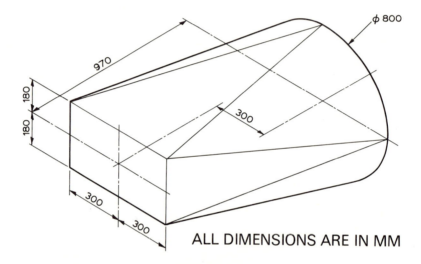

ALL DIMENSIONS ARE IN MM

FIG. 7.18

SCALE 1/20

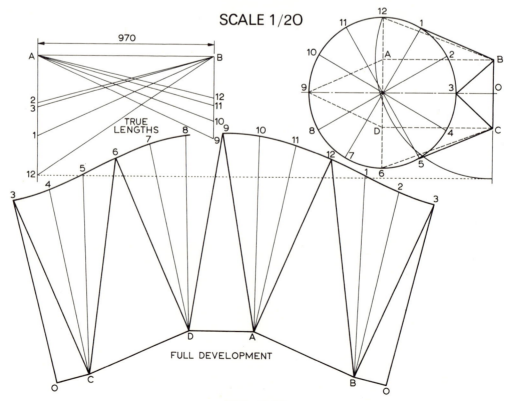

FULL DEVELOPMENT

FIG. 7.19

Triangulation

The oblique cone was developed by a process of triangulation. This is a process in which the true lengths of two intersecting lines, on the solid surface, are determined and drawn out as two sides of a plane triangle; the third side comes from a convenient short line that lies, as nearly as possible, on the solid surface. The same process is particularly useful for transition pieces, such as the one shown in *Figure 7.17*, where a change from a polygonal section duct to a closed-curve section duct is involved.

Views, as seen in the directions of arrows X and Z, have been drawn in orthographic projection. The true length of a line, such as P-5, has been constructed by the same method that was used for an oblique cone. The complete development would consist of four isosceles triangles (one of which appears in its true shape in the view in the direction of arrow X) and four shapes, like P-4-7, which has been constructed by treating the curved surface as three adjacent triangles. Only half of the development is shown in Fig.7.17.

In general, the true length of a line is constructed by drawing, in effect, an auxiliary view as seen in a direction perpendicular to the line, as it appears in one of the views of the orthographic drawing. Of course, the auxiliary view need not be indicated and labelled as shown in Chapter 6; it is only a step in the construction of the development. Nevertheless, it is a convenient way to work out how to determine a true length; the construction shown in Fig.7.17 is based on the principle illustrated in Fig.6.30, where the first auxiliary view shows the true length of line P_1-P_2.

EXERCISE 7.3

Figure 7.18 is a pictorial view of a transition piece for a wind tunnel, in which the section changes from a rectangle to a circle. N.B. The rectangle and the circle are <u>not</u> coaxial.

Using a suitable scale, construct a development of the surface of the transition piece. When you have finished, compare your solution with *Figure 7.19*.

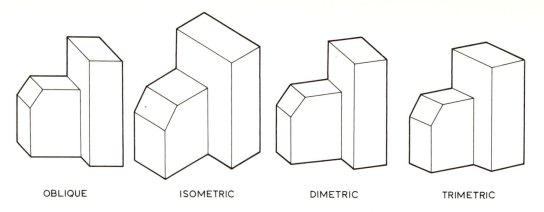

OBLIQUE   ISOMETRIC   DIMETRIC   TRIMETRIC

FIG. 8.1

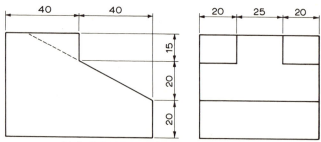

ALL DIMENSIONS ARE IN MM
THIRD ANGLE PROJECTION

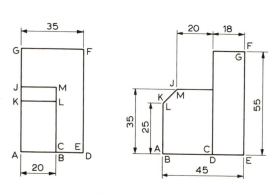

THIRD ANGLE PROJECTION
ALL DIMENSIONS ARE IN MM

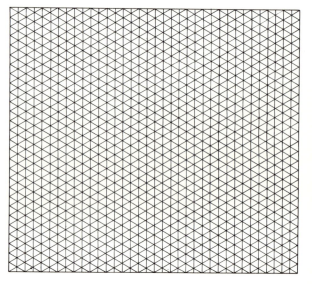

FIG. 8.3

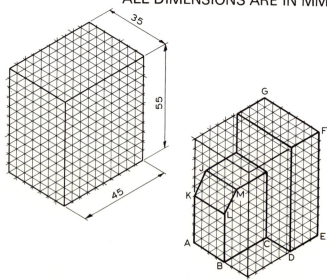

FIG. 8.2

CHAPTER 8

Pictorial Drawings

Previous Chapters have been mainly concerned with orthographic projection: a drawing system that is particularly suitable for detail and general arrangement drawings that have to provide precise manufacturing instructions. In design, training and sales work it is often useful to be able to represent objects pictorially without going to the extent of making a proper perspective drawing. *Figure 8.1* shows four well-established methods of doing this with the aid of only the usual drawing equipment:

(1) Oblique drawing
(2) Isometric drawing
(3) Dimetric drawing
(4) Trimetric drawing.

Most people find that the methods in the above list are in order of ease of making the drawing and in reverse order of pleasing results. This Chapter is devoted mainly to isometric drawings as they are not too difficult to produce and have a fairly pleasing appearance.

Isometric projection

Fig.6.34 (p.64) shows the meaning of isometric projection; a view of a cube along the long diagonal shows the object with all its sides having equal length and the angles between them are either 60° or 120°. Fig.6.35 shows how a pictorial view of an object may be produced either by isometric projection or by isometric drawing. In the former case, double auxiliary projection is used and, in the latter, three-dimensional Cartesian coordinates are transferred to three coplanar axes with 120° between them.

Isometric drawing

Figure 8.2 is the same object as Fig.6.35; the isometric drawing process is shown in more detail. The isometric grid, in the centre, is ruled out with vertical lines and lines at 60° either side of the vertical. Each line intersects the other lines at intervals of 5 mm. The object would fit into a rectangular block 55 mm high, 45 mm long and 35 mm wide. This block has been drawn out on the grid.

On the lower grid, all vertical heights have been transferred to the vertical set of lines, measuring from the lower horizontal plane, ABCDE. All lengths have been transferred to the set of lines inclined at 60° clockwise to the vertical, measuring from the vertical plane EFG. All widths have been transferred to the set of lines inclined at 60° anticlockwise to the vertical, measuring from the vertical plane containing G and A.

It is convenient to think of this process as a method of plotting out POINTS. It is always possible to read or to construct the Cartesian coordinates of all the points contained in a complete detail drawing. After these points have been plotted on the isometric

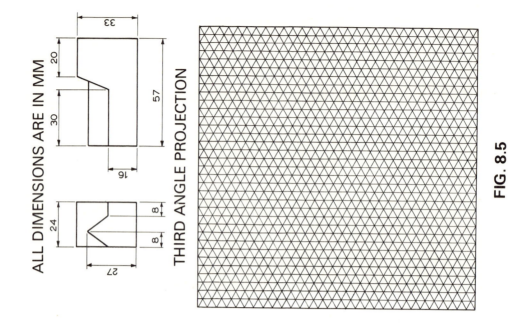

ALL DIMENSIONS ARE IN MM

THIRD ANGLE PROJECTION

FIG. 8.5

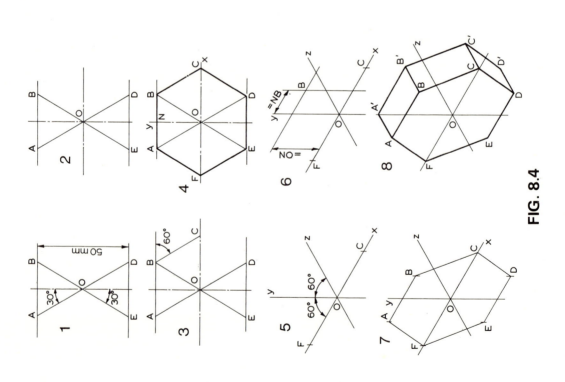

FIG. 8.4

grid it is possible to decide the correct ways of drawing in the lines.

Sloping surfaces and curves often cause some difficulties. Sloping surfaces are plotted out by locating the points that form the plane of the surface, e.g. points J, K, L, M in Fig.8.2. Sometimes the coordinates of these points are known from the dimensions given on the drawing but it is often necessary to construct (or calculate) part of the geometry of the object in order to determine the necessary dimensions.

Curves can be dealt with by treating them as a finite set of points; each point has to be plotted on the isometric grid. The set of points is then joined up by careful sketching or the use of french curves or a flexicurve. Circular arcs and complete circles that lie in one of the isometric planes may be dealt with by special techniques described below.

It is not essential to have an isometric grid in order to make an isometric drawing: a 60° set-square and a scale may be used with ordinary drawing paper. However, for quick isometric sketches and for learning the technique of isometric drawing, a grid is very useful: pads of isometric grid paper are on the market.

The object appearing in the lower grid of Fig.8.2 looks larger than it does in the orthographic drawing at the top of that figure. The actual dimensions were used along the isometric axes whereas any line which is not viewed along a line perpendicular to itself should be foreshortened. It was shown, in Chapter 6, that the object has been magnified by a factor of $\sqrt{3}/\sqrt{2}$. It would be possible to correct for this by having an isometric scale incorporated in the grid, e.g. 2 divisions would represent 10 mm but the actual distance between the divisions would be $10\sqrt{2}/\sqrt{3}$ mm. In most practical applications there is no need to correct for the foreshortening effect, as it affects the axes equally. The slight magnification that results from the use of actual dimensions can be accepted.

EXERCISE 8.1

Use the grid provided to construct an isometric drawing of the object shown in *Figure 8.3*. Choose the view carefully so that as much of the detail of the object is revealed as possible. Do not use the isometric scale ($\sqrt{2}/\sqrt{3}$) and do not add dimensions. Make the visible lines much bolder than the grid lines.

Sloping surfaces

In Exercise 8.1, it is fairly easy to locate all the eight points lying on the sloping surface, but the following example shows how it is often necessary to construct some of the geometry of the object before commencing the isometric drawing. Consider a regular hexagon of 50 mm across flats, cut from a plate that is 15 mm thick. *Figure 8.4* shows the steps in the construction of an isometric drawing of such an object. In steps 1 to 4, a scale drawing of the hexagonal shape has been made. Steps 5 to 8 are stages in the construction of the isometric drawing itself. The steps are as follows:

(1) Draw two parallel lines, AB and ED, 50 mm apart, to subtend an angle of 60° at a point O, midway between them.
(2) Draw a line through O, parallel to AB and ED.
(3) Locate point C by drawing a line through B at 60° to the horizontal.
(4) Locate the sixth point F by drawing a line through A at 60° to the horizontal. Complete the outline of the hexagon. Choose axes Ox and Oy.
(5) Construct isometric axes Ox, Oy and Oz at the correct angles for the isometric drawing. Let point O correspond to the point O at the centre of the hexagon. Points F and C lie on the Ox axis and distances OC and OF are made equal in the drawings for step (4) and step (5), i.e. the isometric scale is not used.
(6) Locate point B as shown, using distances obtained from the drawing for step (4). On the isometric drawing, distances are always measured parallel to either Ox, Oy or Oz.
(7) Locate A, E and D in exactly the same way that B was located. Complete the hexagon.

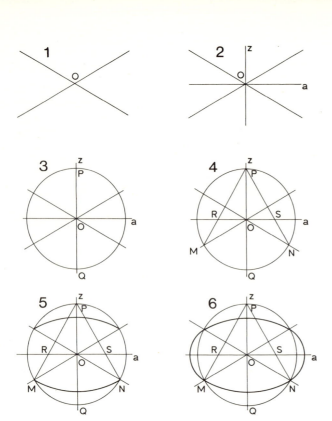

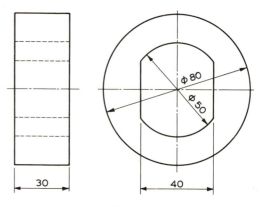

FIG. 8.8

THIRD ANGLE PROJECTION
ALL DIMENSIONS ARE IN MM

FIG. 8.6

THIRD ANGLE PROJECTION

ALL DIMENSIONS ARE IN MM

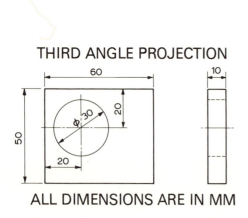

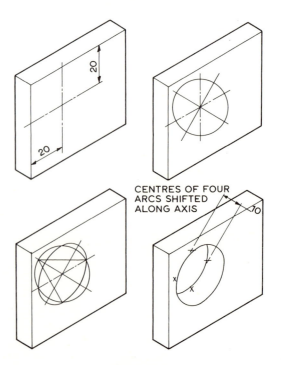

FIG. 8.7

(8) Locate the visible points at the other end of the plate by constructing lines parallel to Oz.
AA' = BB' = CC' = DD' = 15 mm.
Complete the drawing.

The need to produce a scale drawing of part of an object before commencing the isometric drawing can sometimes be avoided when the nature of isometric drawing is fully understood. However, the beginner would be well advised to work from scale drawings at first.

EXERCISE 8.2

Use the grid provided to produce an isometric drawing of the object shown in *Figure 8.5*. The viewing direction should reveal as much detail as possible and, without using the isometric scale, the object should appear about twice full size. Do not dimension.

Circles

Any circle may be treated as a finite number of points and these can be plotted out on the isometric drawing. The set of points is then joined with a smooth curve. This process is very time-consuming and is not always necessary, since the isometric drawing usually serves as an illustration, and only a rough approximation to an ellipse is required. (Strictly speaking, when the isometric scale is not used, a circle of diameter d which lies in one of the isometric planes will appear as an ellipse with major axis $d\sqrt{3}/\sqrt{2}$ and minor axis $d/\sqrt{2}$.)

A very useful approximate method for isometric circles is known as the "Four Arc" method. This is described below and shown in *Figure 8.6*; the isometric scale is not used.

(1) Locate the centre O of the isometric circle by drawing two lines parallel to the isometric axes that define the plane in which the circle lies, and in the appropriate position for the centre of the circle.

(2) Draw two perpendicular lines through O to bisect the isometric axes for the plane in which the circle lies. One of these lines will be the third isometric axis, Oz.

(3) Lightly draw a circle, centre O, radius equal to that of the required isometric circle, to intersect Oz at P and Q.

(4) Draw two lines through P (or Q) at 30° to Oz which cut the circle at M and N on the isometric axes after intersecting Oa at R and S.

(5) Construct two arcs, centres P and Q, radius PM which are bounded by the circle.

(6) Construct two arcs, centres R and S, radius RM which are bounded by the circle.

(7) The four arcs produce a shape which closely follows the correct ellipse. All construction lines have been removed.

(8) The same construction is shown for the other two isometric planes.

Figure 8.7 shows the same process as some of the steps in the making of an isometric drawing of a rectangular plate with a circular hole. Notice how the centres of some of the four arcs have been displaced along an isometric axis to facilitate the drawing of an isometric circle in a plane parallel to the first one. The Four Arc method requires a lot of construction lines; they should be drawn very faintly for easy erasure.

EXERCISE 8.3

Using a drawing board and instruments, make an isometric drawing of the component shown in *Figure 8.8*. Without using the isometric scale, the component should appear about full size. Hint: It is possible to use the Four Arc method for the hole as well as the complete circles; it can be treated as a complete circle cut by two parallel lines.

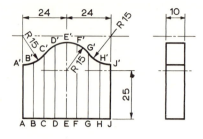

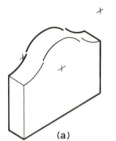

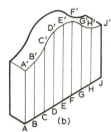

(a)

(b)

Four arc method
gives curves that
do not blend

Along the isometric
axes the distances
between letters equal
those in orthographic
drawing

FIG. 8.9

THIRD ANGLE PROJECTION
ALL DIMENSIONS ARE IN MM

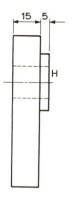

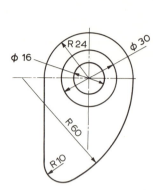

FIG. 8.10

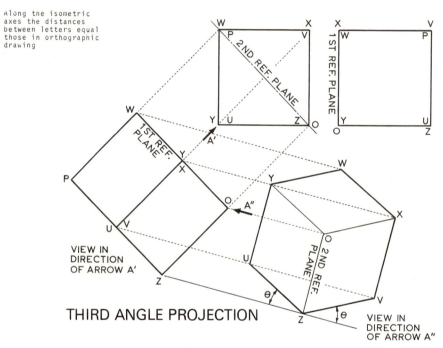

VIEW IN
DIRECTION
OF ARROW A'

VIEW IN
DIRECTION
OF ARROW A"

THIRD ANGLE PROJECTION

FIG. 8.12

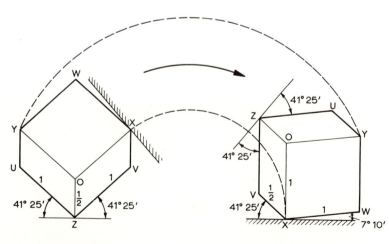

FIG. 8.13.1

FIG. 8.13.2

Blended arcs

The four arc method cannot be used when a number of circular arcs blend together (with a common normal at the transition point). *Figure 8.9* shows such a system and an attempt to use the four arc method. Because the ellipses are not reproduced accurately, especially at the ends of the major axes, the arcs will not blend. It is sometimes possible to manipulate the arcs and achieve satisfactory blending but, as in the successful pictorial drawing in Fig.8.9, it may be just as quick to plot the arcs out as a finite number of points.

EXERCISE 8.4

Make an isometric drawing, about twice full size, of the cam shown in *Figure 8.10*. The surface marked H is to be visible. Use the four arc method only where it is appropriate.

Small circles

Many engineering components have fillet radii and other small curves that would take a great deal of time to construct in an isometric drawing. Small circles (with a radius of less than about 5 mm) may be drawn by locating the four points which lie on the isometric axes and sketching an ellipse through these four points. A satisfactory ellipse can be sketched, provided the axes of symmetry are taken into account. It may be seen in Fig.8.6 that the axes of symmetry always bisect the isometric axes for the plane in which the circle lies.

EXERCISE 8.5

Make an isometric drawing, about full size, of the component shown in *Figure 8.11*. Show as much detail as possible by carefully selecting the viewing direction.

Dimetric projection

In Chapter 6, Fig.6.34, it was shown that an isometric view is a view along the long diagonal, OP, of a cube; all the edges are equally foreshortened. In *Figure 8.12*, the first auxiliary view is in the same direction as that in Fig.6.34 but the second auxiliary view is not along OP. The result is that only two of three intersecting edges are equally foreshortened: this is known as a Dimetric View. By varying the direction of arrow A", the angle θ can be changed; this affects the ratio of the lengths of the edges UZ and VZ to the length of OZ.

Dimetric drawing

A dimetric drawing can be made directly, in the same way as an isometric drawing, but the process is more complicated. In an isometric drawing the sides of a cube are in the ratio 1:1:1 and the angles between the vertical sides and the horizontal sides are both 60°. *Figure 8.13.1* shows the correct angles for a dimetric drawing of a cube in which the vertical sides are half the length of the horizontal sides. This view is far from pleasing, but it can be reorientated, as in *Figure 8.13.2*, so that one of the longer sides is vertical; the appearance is greatly improved.

Figure 8.14 shows how the object illustrated in Fig.8.1 is drawn as a dimetric view with ratios 1:1:½. The correct angles for other ratios can be determined by an analysis of the geometry of Fig.8.12.

EXERCISE 8.6

Make a dimetric drawing of the object shown in Fig.8.5, using ratios of 1:1:½. Make your viewing direction about the same as the solution to Exercise 8.2 and compare the two pictorial views.

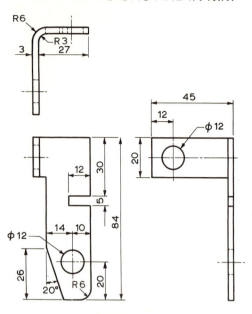

FIG. 8.11

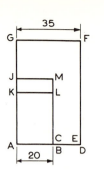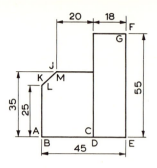

THIRD ANGLE PROJECTION
ALL DIMENSIONS ARE IN MM

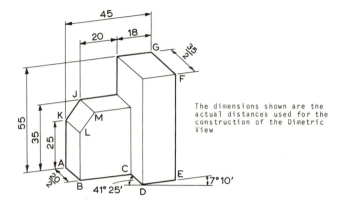

The dimensions shown are the actual distances used for the construction of the Dimetric View

FIG. 8.14

THIRD ANGLE PROJECTION

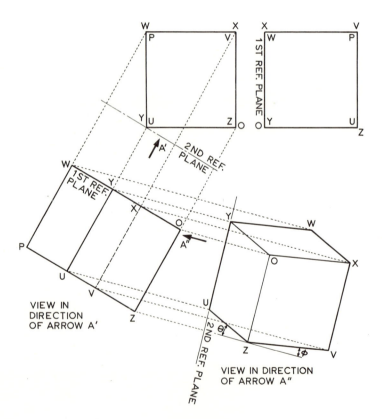

VIEW IN DIRECTION OF ARROW A'

VIEW IN DIRECTION OF ARROW A"

FIG. 8.15

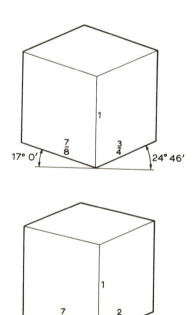

FIG. 8.16

The correct angles to be used with convenient ratios may be calculated from:

$$\left(\frac{OX}{OZ}\right)^2 + \left(\frac{OY}{OZ}\right)^2 = \frac{1 + \tan\theta\,\tan\phi}{1 - \tan\theta\,\tan\phi} \quad \text{and}$$

$$\left(\frac{OX}{OZ}\right)^2 - \left(\frac{OY}{OZ}\right)^2 = \frac{\tan^2\theta - \tan^2\phi}{\tan^2\theta + \tan^2\phi}$$

The dimensions shown are the actual distances used
for the construction of the Trimetric View.

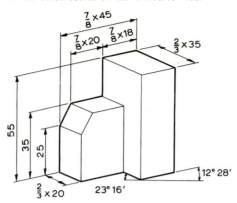

FIG. 8.17

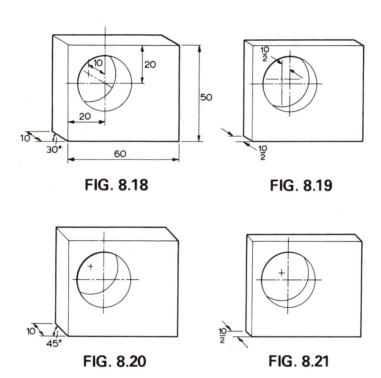

FIG. 8.18   **FIG. 8.19**

FIG. 8.20   **FIG. 8.21**

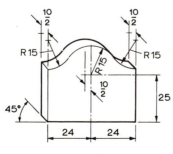

FIG. 8.22.1

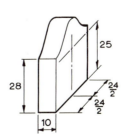

FIG. 8.22.2

Trimetric projection

A double auxiliary view of a cube that shows unequal foreshortening of three intersecting edges, is known as a Trimetric View. *Figure 8.15* shows how both arrows A' and A" are in different directions from the corresponding arrows in Fig.6.34. There is a unique relationship between the angles θ and φ and the ratio of the lengths UZ:VZ:OZ.

Trimetric drawing

Figure 8.16 shows the correct angles for two trimetric views of a cube that have convenient ratios. *Figure 8.17* shows how the trimetric view contained in Fig.8.1 was constructed.

Methods of producing dimetric and trimetric drawings

The construction of these pictorial views may be facilitated by the use of:

(1) Grid paper, like that used for isometric drawings, with the lines ruled at the appropriate angles and spaced according to the corresponding ratios.
(2) A special template scale having slots at the appropriate angles, the edges of the slots being graduated according to the ratios of the dimetric/trimetric scale.

Circles are troublesome in both dimetric and trimetric drawings: this may account for the lack of popularity of these methods.

EXERCISE 8.7

Make a trimetric drawing of the object shown in Fig.8.5, using either of the sets of ratios shown in Fig. 8.16. Make your viewing direction about the same as the solution to Exercise 8.6 and compare the two pictorial views.

Oblique drawing

The great advantage of oblique drawing is that one set of parallel planes of the object appears in its true shape. If the planes concerned contain circular features then the construction of the pictorial view becomes very much easier than for any of the other three methods described in this Chapter.

Figure 8.18 shows an oblique drawing of the same object which appears in Fig.8.7. The angles used are 0° and 30° and the ratios are 1:1:1. It is not possible to produce such a view by double auxiliary projection; an oblique view is simply a convenient, fictitious pictorial representation of the object.

The appearance of an oblique drawing may be improved by changing the ratios to 1:1:½; this has been done in *Figure 8.19*. Moreover, it is possible to use almost any angle for the oblique axis. *Figure 8.20* has ratios 1:1:1 and the oblique axis is at 45°; this is known as a "Cavalier" oblique drawing. *Figure 8.21* has ratios 1:1:½ and the oblique axis is at 45°; this is known as a "Cabinet" oblique drawing.

The oblique drawing contained in Fig.8.1 is also a "Cabinet" oblique drawing.

Figure 8.22.1 illustrates the correct use of oblique projection where drawing time is saved by showing the blended arcs (from Fig.8.9) in their true shape. There are two objections to *Figure 8.22.2*: it is more difficult to draw and the result looks rather peculiar.

EXERCISE 8.8

Make an oblique drawing of the object in Fig.8.8. Use any suitable angle and ratios.

EXERCISE 8.9

Make a "Cabinet" oblique drawing of the cam shown in Fig.8.10. Show the surface labelled H.

CHAPTER 9

Mating Parts

The effects of small differences between the size stated on a drawing and the manufactured size are briefly examined in Chapter 5. In practice it is always necessary to allow a "tolerance" on every dimension: even if it were possible to manufacture with complete accuracy and to verify this by perfect measurement, the cost would be uneconomic. The "tolerance" is the difference between the maximum acceptable size and the minimum acceptable size.

The correct functioning of an assembly of parts is the responsibility of the design department of an organization. It follows that tolerances should be selected by the design personnel rather than those responsible for the manufacturing processes.

A tolerance will depend on whether and how a feature from one part is to fit into (or "mate" with) another part. *Figure 9.1* is a set of detail drawings of four of the five parts shown in Fig. 5.14 (p.42). Clearly, the dimensions *A* and *B* on the pin do not affect the functioning of the assembly and a general tolerance, say 1·0 mm, might be acceptable. However, the dimensions *X* and *Y* do affect the functioning and greater accuracy is required. Furthermore, it is important to ensure that the pin can be inserted in the hole, i.e. the effect of the tolerances on the diameters of the pin and the hole is that the former does not significantly exceed the latter; a slight interference or "push" fit might be acceptable. The nature of the fit depends upon the "deviation" or difference between the actual size and the basic size" (10 mm) of both the pin and the hole.

Tolerances, deviations and limits

Figure 9.2 illustrates the meanings of the terms "tolerance", "deviation" and "limit". For convenience, any two mating parts will be referred to as a shaft and a hole regardless of their actual form. A <u>positive</u> deviation means that the size is <u>larger</u> than the basic size. Thus, for the dimension shown in *Figure 9.3*,

the basic size is 10 mm,

the upper deviation is + 0·006 mm,

the lower deviation is – 0·005 mm,

the tolerance is + 0·006 – (–0·005)
$$= 0·011 \text{ mm},$$

the maximum limit of size is
$$10·000 + 0·006 = 10·006 \text{ mm},$$

the minimum limit of size is
$$10·000 - 0·005 = 9·995 \text{ mm}.$$

The location of the limits of size with respect to the basic size is expressed in terms of the "fundamental" deviation. The deviation that relates to the limit which is <u>nearer</u> to the basic size is termed the "fundamental" deviation. Most systems of limits and fits are based on standardized tolerances and standardized fundamental deviations.

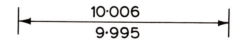

FIG. 9.3

THIRD ANGLE PROJECTION

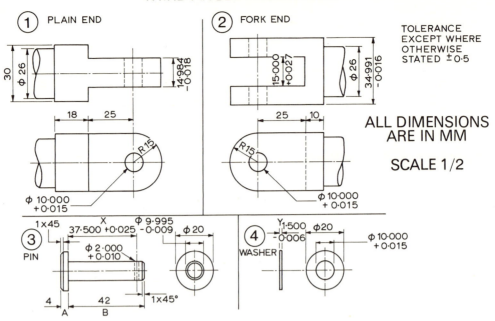

① PLAIN END

30
Φ 26

14·984
−0·018

18   25

R 15

Φ 10·000
+0·015

② FORK END

15·000
0
−0·027

Φ 26

34·991
0
−0·016

25   10

R 15

Φ 10·000
+ 0·015

TOLERANCE
EXCEPT WHERE
OTHERWISE
STATED ±0·5

ALL DIMENSIONS
ARE IN MM

SCALE 1/2

③ PIN

1×45

X   Φ 9·995
37·500 +0·025   −0·009

Φ 2·000
+ 0·010

Φ 20

1×45°

4   42

A   B

④ WASHER

Y 1·500
−0·006

Φ 20

Φ 10·000
+ 0·015

FIG. 9.1

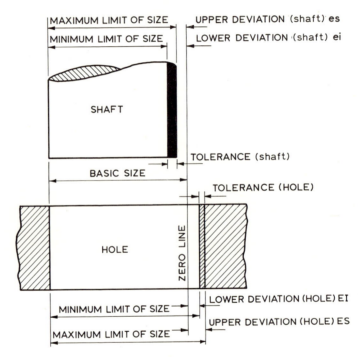

MAXIMUM LIMIT OF SIZE

MINIMUM LIMIT OF SIZE

UPPER DEVIATION (shaft) es

LOWER DEVIATION (shaft) ei

SHAFT

TOLERANCE (shaft)

BASIC SIZE

TOLERANCE (HOLE)

HOLE

ZERO LINE

MINIMUM LIMIT OF SIZE

MAXIMUM LIMIT OF SIZE

LOWER DEVIATION (HOLE) EI

UPPER DEVIATION (HOLE) ES

FIG. 9.2

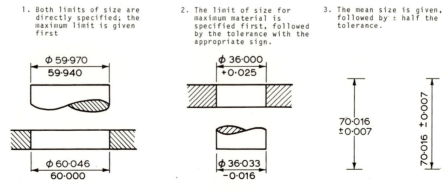

1. Both limits of size are directly specified; the maximum limit is given first

Φ 59·970
59·940

Φ 60·046
60·000

2. The limit of size for maximum material is specified first, followed by the tolerance with the appropriate sign.

Φ 36·000
+0·025

Φ 36·033
−0·016

3. The mean size is given, followed by ± half the tolerance.

70·016
±0·007

70·016 ±0·007

FIG. 9.4

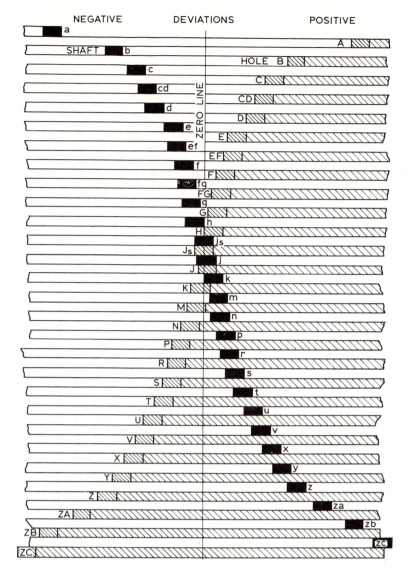

FIG. 9.5

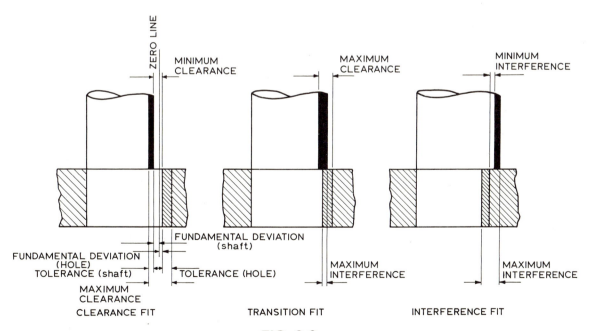

FIG. 9.6

FIG. 9.7

BRITISH STANDARD

SELECTED ISO FITS : HOLE BASIS

ES = Upper deviation (HOLE)   EI = Lower deviation (HOLE)

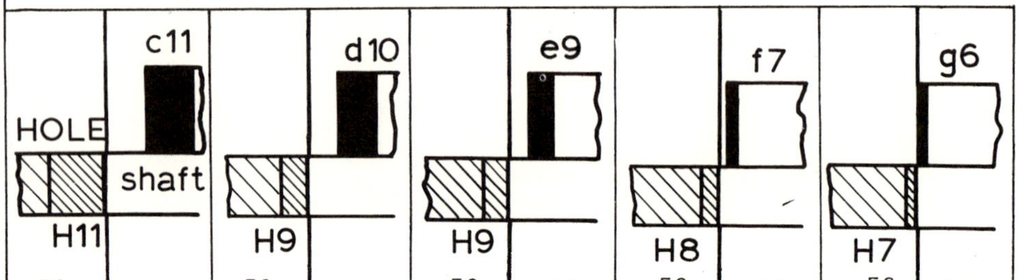

Clearance fits

Over	To	H11 (ES EI)	c11 (es ei)	H9 (ES EI)	d10 (es ei)	H9 (ES EI)	e9 (es ei)	H8 (ES EI)	f7 (es ei)	H7 (ES EI)	g6 (es ei)
mm	mm	0·001 mm	0·001 mm	0·001 mm	0·001 mm	0·001 mm	0·001 mm	0·001 mm	0·001 mm	0·001 mm	0·001 mm
—	3	+ 60 / 0	− 60 / − 120	+ 25 / 0	− 20 / − 60	+ 25 / 0	− 14 / − 39	+ 14 / 0	− 6 / − 16	+ 10 / 0	− 2 / − 8
3	6	+ 75 / 0	− 70 / − 145	+ 30 / 0	− 30 / − 78	+ 30 / 0	− 20 / − 50	+ 18 / 0	− 10 / − 22	+ 12 / 0	− 4 / − 12
6	10	+ 90 / 0	− 80 / − 170	+ 36 / 0	− 40 / − 98	+ 36 / 0	− 25 / − 61	+ 22 / 0	− 13 / − 28	+ 15 / 0	− 5 / − 14
10	18	+ 110 / 0	− 95 / − 205	+ 43 / 0	− 50 / − 120	+ 43 / 0	− 32 / − 75	+ 27 / 0	− 16 / − 34	+ 18 / 0	− 6 / − 17
18	30	+ 130 / 0	− 110 / − 240	+ 52 / 0	− 65 / − 149	+ 52 / 0	− 40 / − 92	+ 33 / 0	− 20 / − 41	+ 21 / 0	− 7 / − 20
30	40	+ 160 / 0	− 120 / − 280	+ 62 / 0	− 80 / − 180	+ 62 / 0	− 50 / − 112	+ 39 / 0	− 25 / − 50	+ 25 / 0	− 9 / − 25
40	50	+ 160 / 0	− 130 / − 290								
50	65	+ 190 / 0	− 140 / − 330	+ 74 / 0	− 100 / − 220	+ 74 / 0	− 60 / − 134	+ 46 / 0	− 30 / − 60	+ 30 / 0	− 10 / − 29
65	80	+ 190 / 0	− 150 / − 340								
80	100	+ 220 / 0	− 170 / − 390	+ 87 / 0	− 120 / − 260	+ 87 / 0	− 72 / − 159	+ 54 / 0	− 36 / − 71	+ 35 / 0	− 12 / − 34
100	120	+ 220 / 0	− 180 / − 400								
120	140	+ 250 / 0	− 200 / − 450	+ 100 / 0	− 145 / − 305	+ 100 / 0	− 84 / − 185	+ 63 / 0	− 43 / − 83	+ 40 / 0	− 14 / − 39
140	160	+ 250 / 0	− 210 / − 460								
160	180	+ 250 / 0	− 230 / − 480								
180	200	+ 290 / 0	− 240 / − 530	+ 115 / 0	− 170 / − 355	+ 115 / 0	− 100 / − 215	+ 72 / 0	− 50 / − 96	+ 46 / 0	− 15 / − 44
200	225	+ 290 / 0	− 260 / − 550								
225	250	+ 290 / 0	− 280 / − 570								
250	280	+ 320 / 0	− 300 / − 620	+ 130 / 0	− 190 / − 400	+ 130 / 0	− 110 / − 240	+ 81 / 0	− 56 / − 108	+ 52 / 0	− 17 / − 49
280	315	+ 320 / 0	− 330 / − 650								
315	355	+ 360 / 0	− 360 / − 720	+ 140 / 0	− 210 / − 440	+ 140 / 0	− 125 / − 265	+ 89 / 0	− 62 / − 119	+ 57 / 0	− 18 / − 54
355	400	+ 360 / 0	− 400 / − 760								
400	450	+ 400 / 0	− 440 / − 840	+ 155 / 0	− 230 / − 480	+ 155 / 0	− 135 / − 290	+ 97 / 0	− 68 / − 131	+ 63 / 0	− 20 / − 60
450	500	+ 400 / 0	− 480 / − 880								

es = Upper deviation (shaft)   ei = Lower deviation (shaft)

Transition fits						Interference fits					
h6		k6		n6		p6		s6			
H7 ES EI	es ei	H7 ES EI	es ei	H7 ES EI	es ei	H7 ES EI	es ei	H7 ES EI	es ei		
H7	**h6**	**H7**	**k6**	**H7**	**n6**	**H7**	**p6**	**H7**	**s6**	**Over**	**To**
0·001 mm	0·001 mm	0·001 mm	0·001 mm	0·001 mm	0·001 mm	0·001 mm	0·001 mm	0·001 mm	0·001 mm	mm	mm
+10 / 0	−6 / 0	+10 / 0	+6 / +0	+10 / 0	+10 / +4	+10 / 0	+12 / +6	+10 / 0	+20 / +14	—	3
+12 / 0	−8 / 0	+12 / 0	+9 / +1	+12 / 0	+16 / +8	+12 / 0	+20 / +12	+12 / 0	+27 / +19	3	6
+15 / 0	−9 / 0	+15 / 0	+10 / +1	+15 / 0	+19 / +10	+15 / 0	+24 / +15	+15 / 0	+32 / +23	6	10
+18 / 0	−11 / 0	+18 / 0	+12 / +1	+18 / 0	+23 / +12	+18 / 0	+29 / +18	+18 / 0	+39 / +28	10	18
+21 / 0	−13 / 0	+21 / 0	+15 / +2	+21 / 0	+28 / +15	+21 / 0	+35 / +22	+21 / 0	+48 / +35	18	30
+25 / 0	−16 / 0	+25 / 0	+18 / +2	+25 / 0	+33 / +17	+25 / 0	+42 / +26	+25 / 0	+59 / +43	30	40
										40	50
+30 / 0	−19 / 0	+30 / 0	+21 / +2	+30 / 0	+39 / +20	+30 / 0	+51 / +32	+30 / 0	+72 / +53	50	65
								+30 / 0	+78 / +59	65	80
+35 / 0	−22 / 0	+35 / 0	+25 / +3	+35 / 0	+45 / +23	+35 / 0	+59 / +37	+35 / 0	+93 / +71	80	100
								+35 / 0	+101 / +79	100	120
+40 / 0	−25 / 0	+40 / 0	+28 / +3	+40 / 0	+52 / +27	+40 / 0	+68 / +43	+40 / 0	+117 / +92	120	140
								+40 / 0	+125 / +100	140	160
								+40 / 0	+133 / +108	160	180
+46 / 0	−29 / 0	+46 / 0	+33 / +4	+46 / 0	+60 / +31	+46 / 0	+79 / +50	+46 / 0	+151 / +122	180	200
								+46 / 0	+159 / +130	200	225
								+46 / 0	+169 / +140	225	250
+52 / 0	−32 / 0	+52 / 0	+36 / +4	+52 / 0	+66 / +34	+52 / 0	+88 / +56	+52 / 0	+190 / +158	250	280
								+52 / 0	+202 / +170	280	315
+57 / 0	−36 / 0	+57 / 0	+40 / +4	+57 / 0	+73 / +37	+57 / 0	+98 / +62	+57 / 0	+226 / +190	315	355
								+57 / 0	+244 / +208	355	400
+63 / 0	−40 / 0	+63 / 0	+45 / +5	+63 / 0	+80 / +40	+63 / 0	+108 / +68	+63 / 0	+272 / +232	400	450
								+63 / 0	+292 / +252	450	500

Guide to the selection of fits

Fit	Description	Examples
H11-c11 Slack running fit	Used to give flexibility under load, easy assembly or a close fit at elevated working temperatures.	 12 MM DIA H11-c11 — I.C. engine exhaust valve in guide
H9-d10 Loose running fit	Used for gland seals, loose pulleys and very large bearings.	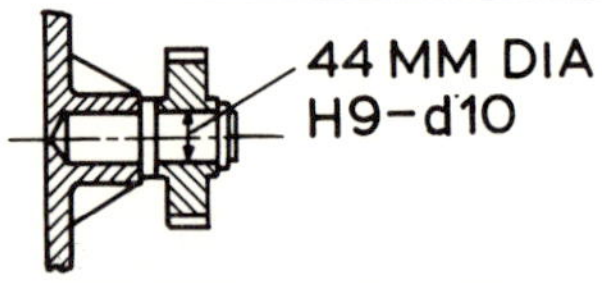 44 MM DIA H9-d10 — Idler gear on spindle
H9-e9 Easy running fit	Used for widely separated bearings or several bearings in line.	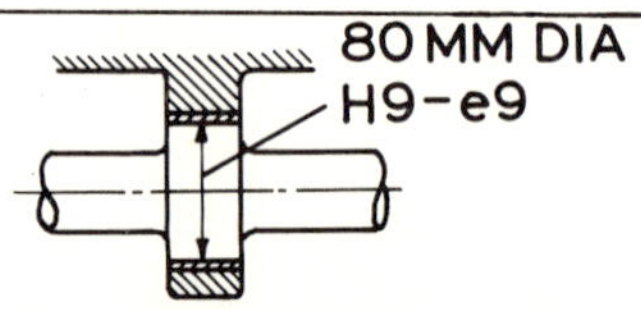 80 MM DIA H9-e9 — Camshaft in bearing
H8-f7 Normal running fit	Suitable for applications requiring a good quality fit that is easy to produce.	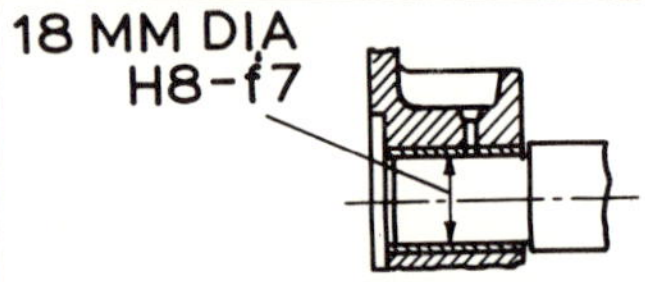 18 MM DIA H8-f7 — Gearbox shaft in bearing
H7-g6 Sliding and location fit	Not normally used for continuously running bearings unless load is slight. Suitable for precision sliding and location.	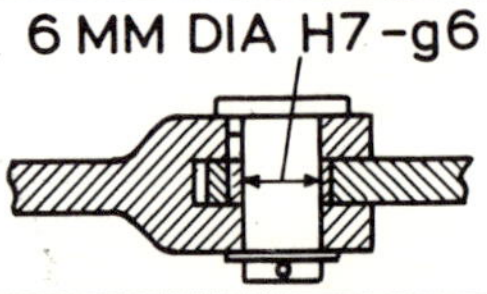 6 MM DIA H7-g6 — Valve mechanism link pin
H7-h6 Location fit	Suitable for many non-running assemblies.	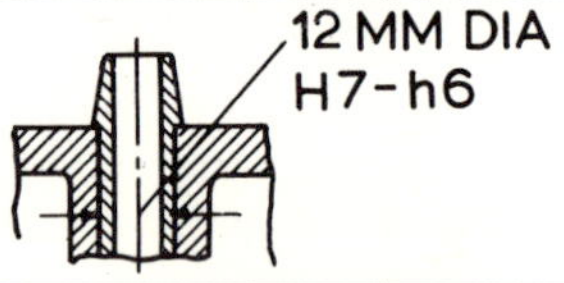 12 MM DIA H7-h6 — Valve guide in head
H7-k6 Push fit	Used for location fits when a slight interference, which eliminates movement of one part relative to tne other, is an advantage.	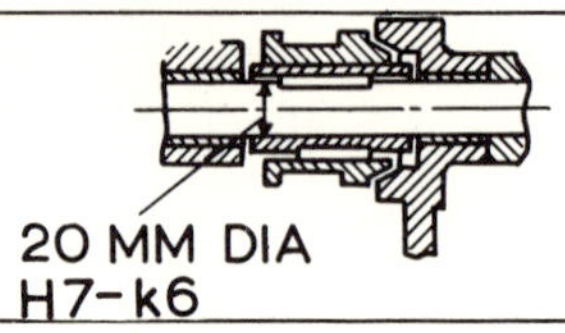 20 MM DIA H7-k6 — Clutch member keyed to shaft
H7-n6 Tight assembly fit	Used when the degree of clearance that can result from a H7-k6 fit is not acceptable.	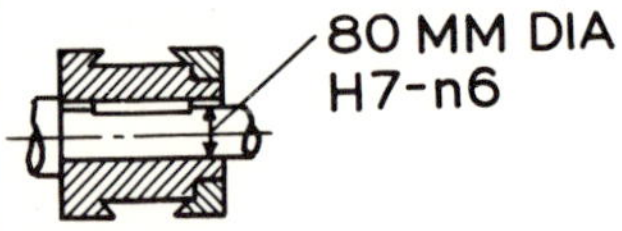 80 MM DIA H7-n6 — Commutator shell on shaft
H7-p6 Press fit	Ferrous parts are not over-strained during assembly or dismantling.	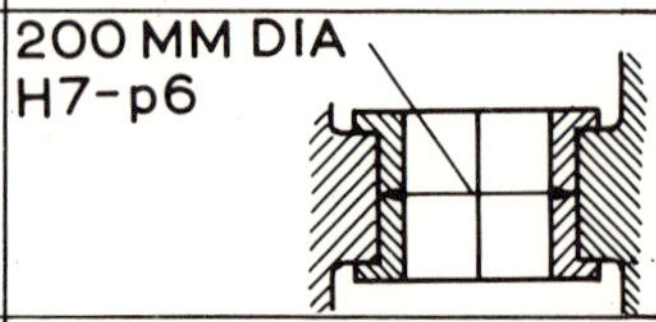 200 MM DIA H7-p6 — Split journal bearing
H7-s6 Heavy press fit	Mainly used for permanent assemblies.	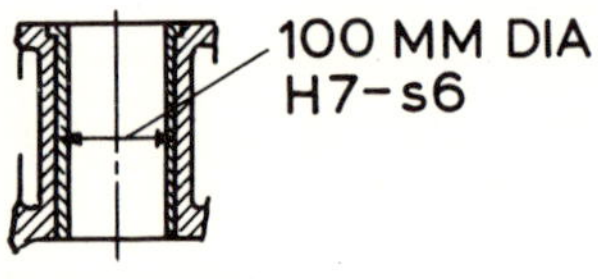 100 MM DIA H7-s6 — Cylinder liner in block

FIG. 9.8

State the upper and lower deviations, the maximum and minimum limits of size, the tolerance and the fundamental deviation for the dimensions shown in *Figure 9.4*. Answers are at the end of the Chapter.

Notice that a dimension may be toleranced in three distinct ways.

Selection of fits

This is a task which requires considerable design experience. However, there are certain applications where the choice of fit is very limited and these, together with an introduction to the International Organization for Standardization (ISO) system of limits and fits, are treated in this Chapter.

British standard 4500:1969

The ISO system has been accepted in Britain, the U.S.A., Canada and many European countries. The system is incorporated in B.S.4500 of 1969 which is wholly metric and replaces the earlier B.S.1916. At first sight, the vast range of tolerances and deviations can be rather bewildering; in fact, the system is very straightforward as there are only three principal variables:

(1) the basic size of the two mating components,
(2) the grade of the tolerance (the smaller the tolerance, the lower the grade) and
(3) the fundamental deviation.

The scope of B.S.4500 is very wide and only a small part of the range of fits would be used for one particular assembly.

The range of basic size extends up to 3 150 mm but, for most applications, a smaller range to 500 mm is sufficient. There are 18 grades of tolerance, designated by the symbols IT01, IT0, IT1, etc., up to IT16. The lower grades are used only for tools and gauges; the upper grades are general tolerances for non-mating parts. In Fig.9.1, the grade of the tolerance on the pin diameter is IT6 and, for the hole in the fork end, it is IT7. There are 28 symbols for fundamental deviations, both for shafts and holes. An upper-case (capital) letter code is used for the holes and a lower case letter code is used for the shafts. *Figure 9.5* illustrates the range of fundamental deviations; the tolerance is represented by the shaded rectangles.

A hole is designated by (using the fork end of Fig.9.1 as an example) 10 mm dia H7; i.e. the basic size is 10 mm, the fundamental deviation is H, and the tolerance grade is IT7. The mating pin (shaft) is 10 mm dia h6. The fit between two mating parts is designated

10 mm dia H7/h6, or 10 mm dia H7-h6.

Figure 9.6 shows the relationships between fundamental deviations, tolerances, clearances and interferences. There are three ways of producing any particular clearance or interference; taking a clearance as the example, the parts could be:

(1) Hole larger than basic size; shaft smaller than basic size.
(2) Hole larger than basic size; maximum limit of shaft equal to basic size. This is termed a unilateral shaft basis system.
(3) Minimum limit of hole equal to basic size; shaft smaller than basic size. This is termed a unilateral hole basis system.

The first system is not practical because of the lack of uniformity. The second system is particularly useful when stock bar material is used for the shaft, or if many components are mounted on a common shaft. The third system is recommended for most other applications as it is usually convenient to make a standard size of hole (with a drill or a reamer) and then produce the shaft to suit it. It may be seen, from Fig.9.5, that holes suitable for a unilateral hole basis system have the letter code H.

Fits for general engineering products

B.S.4500 suggests that a selection of only ten fits, with a unilateral hole basis, will prove suitable for the great majority of applications. These fits are illustrated and the complete list of deviations are given in *Figure 9.7*. The range of basic sizes is up to 500 mm, as anything in excess of this is too large to be considered a general engineering product and special problems arise with thermal expansion.

The industrial user who standardizes on only the ten fits in Fig.9.7 (many users need far less) enjoys a considerable economic and engineering advantage. For any particular basic size, the number of tools and gauges required is reasonably small but, if one of the above fits is not suitable, it is very probable that another combination (without the need of additional gauges) will be acceptable. For example, a H8-f7 fit could sometimes give too much clearance but the H7-g6 fit would be too expensive; from the same stock of gauges, a H7-f7 fit might be just right.

EXERCISE 9.2

Write out toleranced dimensions for Holes and Shafts with the following fits:

100 mm dia H7-h6;   50 mm dia H9-e9;

 20 mm H7-k6;   150 mm H7-s6.

Typical applications of hole basis fits

The ten selected ISO fits are briefly described in *Figure 9.8*. The names given to them are only intended to give some indication of the nature of the fit. The illustrated examples are typical applications but the choice of fit is not automatic. For example, the valve mechanism link pin is shown with a H7-g6 fit with the bush whereas a suitable fit with the fork end could be H7-h6. Unless the assembly is re-designed, the pin cannot change from g6 to h6 deviations along its length. The designer would have to decide whether to accept relative movement between the pin and the fork because of the use of H7-g6 or to eliminate it by using h6 for the pin and a positive fundamental deviation for the hole in the bush.

Remember that Fig.9.8 is only a guide to selection; there are no hard-and-fast rules.

EXERCISE 9.3

Figure 9.9 shows part of a crank-shaft. Which dimensions should be individually toleranced? Select a possible fit for the 50 mm dia shaft end and draw a new, suitably toleranced dimension on the figure.

Other manufacturing tolerances

Apart from errors of size, there are two further classes of error likely to arise in any manufacturing process. A simple example is a hole in a piece of sheet metal; not only can the size of the hole be incorrect but it is also possible for its centre to be in the wrong place (an error of position) and the hole to be non-circular (an error of form). B.S.308 defines a geometrical tolerance as the maximum permissible overall variation of form or position about that shown on the drawing. There are recommended methods of specifying straightness, flatness, parallelism, squareness, angularity, concentricity, symmetry, and position.

It is not always necessary to specify geometrical tolerances as, for example, when the designer is familiar with the quality of components manufactured in his employer's machine shop. When there is a lack of direct contact between the design and manufacturing personnel (sub-contractors, manufacture overseas, etc.) it is worth considering specific geometrical tolerances.

Answers to Exercise 9.1

	1		2		3
	Shaft	Hole	Hole	Shaft	
Upper deviation	−0·030	+0·046	+0·025	+0·033	+0·023
Lower deviation	−0·060	0	0	+0·017	+0·009
Maximum limit of size	59·970	60·046	36·025	36·033	70·023
Minimum limit of size	59·940	60·000	36·000	36·017	70·009
Tolerance	0·030	0·046	0·025	0·016	0·014
Fundamental deviation	−0·030	0	0	+0·017	+0·009

96

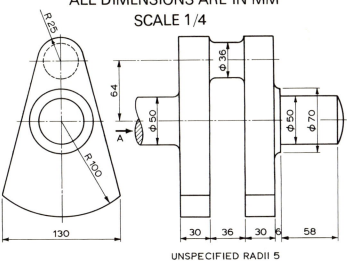

FIRST ANGLE PROJECTION
ALL DIMENSIONS ARE IN MM
SCALE 1/4

R 25
R 100
64
Φ 36
Φ 50
Φ 50
Φ 70
A
130
30 36 30 6 58

UNSPECIFIED RADII 5

FIG 9.9

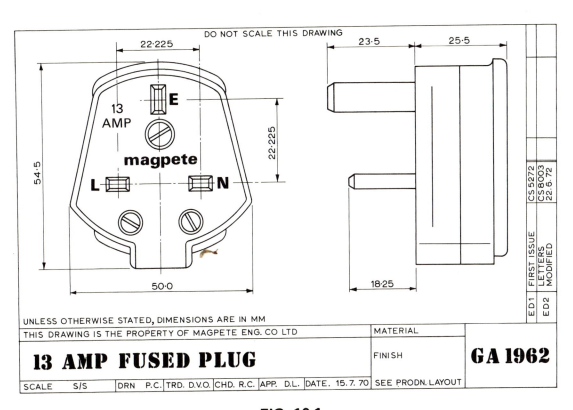

DO NOT SCALE THIS DRAWING

22·225
23·5
25·5

E
13
AMP
22·225
magpete
L
N
54·5

18·25

50·0

CS 5272
CS 8003
22.6.72

FIRST ISSUE
LETTERS
MODIFIED

ED1
ED2

UNLESS OTHERWISE STATED, DIMENSIONS ARE IN MM

THIS DRAWING IS THE PROPERTY OF MAGPETE ENG. CO LTD

	MATERIAL	
13 AMP FUSED PLUG	FINISH	**GA 1962**
SCALE S/S DRN P.C. TRD. D.V.O. CHD. R.C. APP. D.L. DATE. 15.7.70	SEE PRODN. LAYOUT	

FIG. 10.1

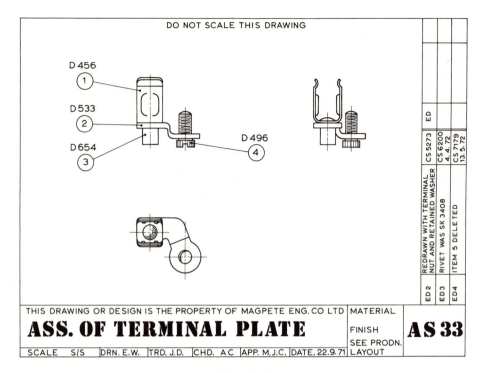

DO NOT SCALE THIS DRAWING

D 456 ①
D 533 ②
D 654 ③
D 496 ④

	ED		
	CS 5273	CS 6200 4.4.72	CS 7179 13.5.72
	REDRAWN WITH TERMINAL NUT AND RETAINED WASHER	RIVET WAS SK 3408	ITEM 5 DELETED
	ED 2	ED3	ED4

THIS DRAWING OR DESIGN IS THE PROPERTY OF MAGPETE ENG. CO LTD

ASS. OF TERMINAL PLATE

MATERIAL
FINISH
SEE PRODN. LAYOUT

AS 33

SCALE S/S | DRN. E.W. | TRD. J.D. | CHD. A C | APP. M.J.C. | DATE. 22.9.71

FIG. 10.2

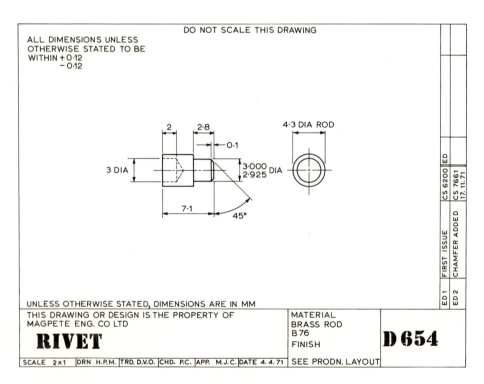

DO NOT SCALE THIS DRAWING

ALL DIMENSIONS UNLESS
OTHERWISE STATED TO BE
WITHIN + 0·12
        − 0·12

2   2·8
0·1
3 DIA
3·000 DIA
2·925
7·1
45°

4·3 DIA ROD

CS 6200	ED	
FIRST ISSUE	CS 7661 17.11.71	
	CHAMFER ADDED	
ED 1	ED 2	

UNLESS OTHERWISE STATED, DIMENSIONS ARE IN MM

THIS DRAWING OR DESIGN IS THE PROPERTY OF
MAGPETE ENG. CO LTD

RIVET

MATERIAL
BRASS ROD
B 76
FINISH
SEE PRODN. LAYOUT

D 654

SCALE 2×1 | DRN H.P.M. | TRD. D.V.O. | CHD. P.C. | APP. M.J.C. | DATE 4.4.71

FIG. 10.3

CHAPTER 10

Engineering Drawings in Industry

Most of the drawings shown in earlier chapters illustrate particular points about engineering drawing: they could not be used in an industrial situation because they lack certain essential features. It has been stated already that the function of an engineering drawing is to facilitate the manufacture of a component, or the correct assembly of a set of components; there is one other important function, namely to illustrate. These three functions are treated in this chapter.

General arrangement drawings

Figure 10.1 is a General Arrangement Drawing of a 13 amp plug: a familiar object that serves to illustrate the differences between each type of drawing. Only certain important dimensions and the major components appear in this drawing; its function is to illustrate the general appearance and basic geometry of the object. A general arrangement drawing does not necessarily show every component; Fig.10.1 gives little indication of, for example, the method by which the cable is to be clamped.

Assembly and sub-assembly drawings

Figure 10.2 shows a sub-assembly of a terminal plate for the 13 amp plug. All the four parts that make up the sub-assembly are shown and each is given an item number (in a circle, from which a leader line indicates the appropriate part and terminates in a dot) and a part number (which often refers to a detail drawing) adjacent to the item number. Some companies have a printed box on their drawing sheets in which the separate items are listed, together with their part numbers and the quantity required for <u>one</u> of the assemblies shown. Other companies prefer to list the items on a separate "Production Layout" or "Bill of Material".

An assembly drawing of the complete plug would have to contain at least one sectional view in order to show all the items. On the assembly drawing of a typical domestic plug there are seventeen items: two of them are sub-assemblies. One sub-assembly has four items and the other has two, so even such a simple object contains twenty-one components and requires twenty-four engineering drawings for manufacture and assembly. This will give the student some indication of the number of drawings that would be required for an internal combustion engine or a machine tool. Much of the cost of an engineering product consists of design and drawing costs; the actual material costs are often very low.

Most drawings are modified in the course of time and the changes are listed on the drawing. The box on the right-hand side of Fig.10.2 has been included for this purpose and many companies have some similar system.

Detail drawings

Figure 10.3 is a detail drawing of Item No. 3 from Fig.10.2. Its function is to provide the appropriate department with precise, complete information for the manufacture of a single component. The drawing number (D654) has been referred to in Fig.10.2. By definition, any detail drawing contains:

1. All necessary views of the object in an identifiable type of orthographic projection.
2. All necessary dimensions.
3. Tolerances.
4. An indication of the material from which the component is to be made.

Depending on the policy of the organization using the drawing it may also contain:

1. Manufacturing instructions.
2. Surface finish instructions.
3. Special treatment instructions.

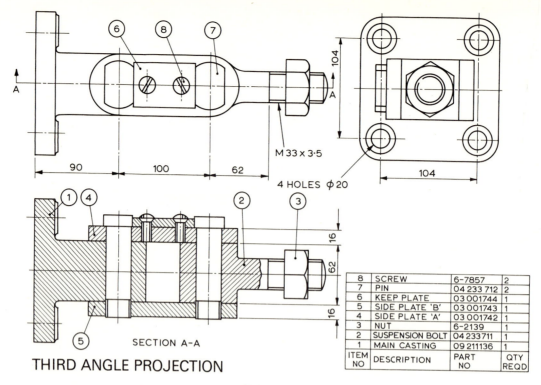

M 33 x 3·5

4 HOLES ⌀20

SECTION A-A

THIRD ANGLE PROJECTION

ITEM NO	DESCRIPTION	PART NO	QTY REQD
8	SCREW	6-7857	2
7	PIN	04 233 712	2
6	KEEP PLATE	03 001744	1
5	SIDE PLATE 'B'	03 001743	1
4	SIDE PLATE 'A'	03 001742	1
3	NUT	6-2139	1
2	SUSPENSION BOLT	04 233711	1
1	MAIN CASTING	09 211136	1

FIG. 10.4

THIRD ANGLE PROJECTION
ALL DIMENSIONS ARE IN MM

SCALE 1/4

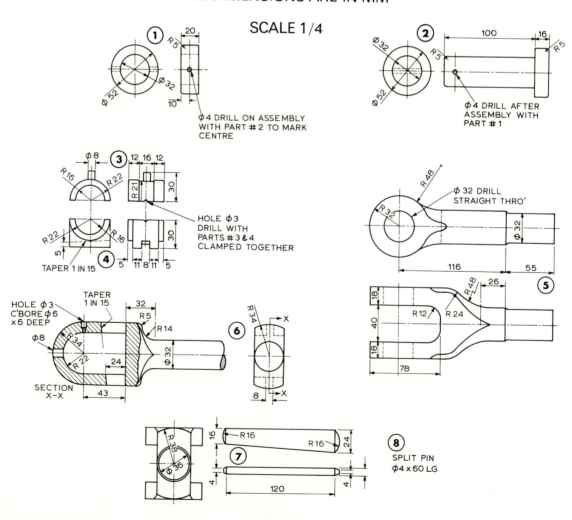

FIG. 10.5

Conventions in assembly drawings

In Chapter 3 (Sections) it was stated that, as well as ribs, there are a number of other common parts that are not shown in section when the section plane passes longitudinally through them. *Figure 10.4* shows three views of an assembly in which a section plane passes longitudinally through two pins and two screws; these fasteners are not shown in section as their internal shape is of no interest to the reader of the drawing, indeed it may be confusing. The other common parts treated in this way are bolts, nuts, rods, rivets, keys, shims and washers.

Fig.10.4 also shows the conventional method of using section lining to distinguish between adjacent parts; the direction and the spacing of the lining is varied. The spacing should take account of the size of the sectioned part; the large main casting has lining with spaces of about twice the width of the small keep plate.

EXERCISE 10.1

Figure 10.5 details eight parts of a heavy duty joint. Part No.3 fits into the 8 mm diameter hole in part No. 6; part No.4 follows it and is wedged into place by part No.7. Part No.5 slides over part No.3 and part No.4 and then part No.2 passes through all of them. Part No.1 fits on the end of part No.2 and is held in place by part No.8.

Draw three views of the complete assembly; one view is to be a section corresponding to section X-X of part No. 6. The other views should be selected to show as much detail as possible. Ignore the intersection lines that appear on parts 5 and 6; they can be indicated by thin straight lines in the form of a V.

Label and list the parts as has been done in Fig.10.4. Show only essential hidden detail. Do not add any dimensions. Insert all necessary titles and labels.

Fasteners

All the parts shown in an assembly or a sub-assembly drawing require either a detail drawing, a designation that identifies a stocked item, or some information about the source of supply. The parts that are most commonly stocked are fasteners with screw threads such as nuts, bolts and screws.

ISO metric threads were briefly introduced in Chapter 5. They are designated as shown in *Figure 10.6*. B.S. 3643 specifies the thread form and two series of diameter-pitch combinations. The two series, one with coarse and the other with fine pitches, are the same as those scheduled in ISO Recommendation R 262. A selection from the full range is shown in *Figure 10.7*; the "First Choice" of basic major diameters (the maximum material diameter of a bolt) are those that would normally be stocked, whilst only larger organizations would stock the full range of "Second Choice" diameters. B.S.3643 also has a list of "Third Choice" diameters.

The various types of fasteners are specified in the following British Standards:

B.S.4190 ISO metric black hexagon bolts, screws and nuts.
B.S.3692 ISO metric precision hexagon bolts, screws and nuts.
B.S.4183 Metric machine screws and machine screw nuts.

The essential difference between a bolt and a screw is that only the latter is threaded right up to the head.

All ISO metric hexagon fasteners are chamfered at 30° to the top of the head and this produces the characteristic shape shown in stage (7) of *Figure 10.8* which is rather awkward to draw. The other views in Fig.10.8 show the steps in the construction. If the hexagon shape appears in one view, it may be constructed as shown in stages (1) and (2). A view showing the distance across the corners is begun in stage (3) and completed, with the aid of an arc template, in stage (4). Height AB = CD = EF = GH. A view showing the distance across the flats is begun in stage (5) by transferring the height AB from stage (3) and completed, again with the aid of an arc template, in stage (6). A bolt with a full bearing head is shown in stage (7); it has been constructed by using only the coefficients for d that are shown in the various stages.

Actual sizes of ISO metric nuts, bolts and screws are given in *Figure 10.9*.

EXERCISE 10.2

Make a general arrangement drawing from the assembly drawing shown in Fig. 10.4. Choose suitable dimensions for sizes that are not shown. Add only major dimensions to the general arrangement drawing.

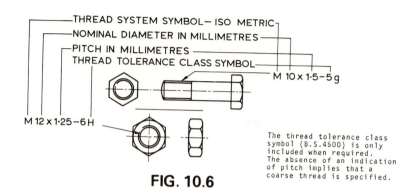

THREAD SYSTEM SYMBOL— ISO METRIC
NOMINAL DIAMETER IN MILLIMETRES
PITCH IN MILLIMETRES
THREAD TOLERANCE CLASS SYMBOL

M 10 x 1·5 – 5g

M 12 x 1·25 – 6H

The thread tolerance class
symbol (B.S.4500) is only
included when required.
The absence of an indication
of pitch implies that a
coarse thread is specified.

FIG. 10.6

FIG. 10.7

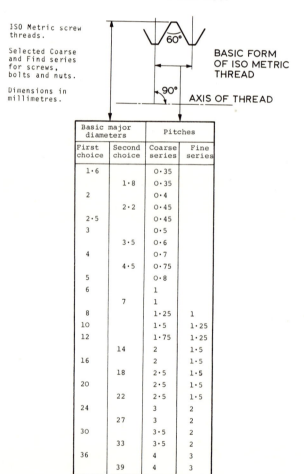

ISO Metric screw
threads.

Selected Coarse
and Find series
for screws,
bolts and nuts.

Dimensions in
millimetres.

BASIC FORM
OF ISO METRIC
THREAD

60°

90° AXIS OF THREAD

Basic major diameters		Pitches	
First choice	Second choice	Coarse series	Fine series
1·6		0·35	
	1·8	0·35	
2		0·4	
	2·2	0·45	
2·5		0·45	
3		0·5	
	3·5	0·6	
4		0·7	
	4·5	0·75	
5		0·8	
6		1	
	7	1	
8		1·25	1
10		1·5	1·25
12		1·75	1·25
	14	2	1·5
16		2	1·5
	18	2·5	1·5
20		2·5	1·5
	22	2·5	1·5
24		3	2
	27	3	2
30		3·5	2
	33	3·5	2
36		4	3
	39	4	3

Surface texture.

Any material surface appears rough when examined under a microscope. Some assemblies will only function correctly if the degree of roughness does not exceed certain limits; these must be specified on the detail drawing. *Figure 10.10* illustrates the recommendations of B.S.308 with regard to symbols for the control of surface finish.

The roughness numbers shown in Fig. 10.10 are explained in B.S.1134.

EXERCISE 10.3

Figure 10.11 shows two views of a cam; surface textures have been indicated but the notes are not clear. There are other errors and omissions in this drawing.

Redraw the two views and insert all dimensions, tolerances and surface texture symbols in accordance with the recommendations of B.S.308. (Refer to Figs. 5.10, 5.11, 5.12, 9.4 and 10.10.)

EXERCISE 10.4

Make detail drawings of items 1 to 7 from Fig.10.4. Choose suitable dimensions for those not shown. Mating sizes should be toleranced after selecting suitable fits from Fig.9.8. Indicate the surfaces that are to be machined but do not show figures for surface texture.

EXERCISE 10.5

Figure 10.12 shows two views of a die casting; some dimensions are missing. After casting, some machining is done to the surfaces that are indicated and certain dimensions are to be given the tolerances that are indicated. Make a new detail drawing of the object in a form that would be suitable for issue to a machine shop, i.e. all the data required for machining and inspection should be present on the drawing.

Modifications to detail drawings

Figs. 10.2 and 10.3 show how modifications are recorded on both assembly and detail drawings. Modifications may be made for many reasons; the component is not strong enough, not rigid enough, is difficult to assemble, wears away too quickly, etc. All too often the task of modifying a detail drawing is given to a junior employee with vague verbal instructions about what is required. If the component gives any further trouble, the diagnostic process is often hampered by poor records of the modification process. Ideally, the revision should be recorded as fully as possible both on the drawing and in the office records and memoranda.

EXERCISE 10.6

Figure 10.13 is a pictorial sketch of a boss for a centrifuge which sheared apart after about 100 hours of running. The component is to be strengthened by the incorporation of six ribs arranged radially on the centre lines of the six holes. The ribs may extend as far as the hole bosses in the radial direction and to within 3 mm of the surface A, in the axial direction. Make a detail drawing of the modified component. List the revision in a suitable revision table, with a succinct description of the changes made. Write a brief note, in the form of a memorandum, describing the modification in more detail.

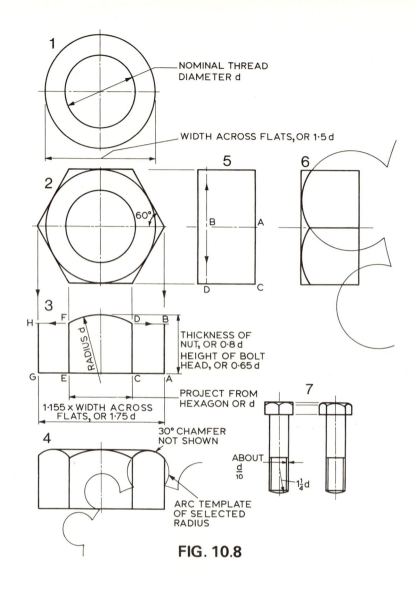

1 — NOMINAL THREAD DIAMETER d

WIDTH ACROSS FLATS, OR 1·5 d

2 — 60°

5 — B / A / D / C

6

3 — H F D B

THICKNESS OF NUT, OR 0·8 d
HEIGHT OF BOLT HEAD, OR 0·65 d

RADIUS d

G E C A

1·155 x WIDTH ACROSS FLATS, OR 1·75 d

PROJECT FROM HEXAGON OR d

30° CHAMFER NOT SHOWN

4

ARC TEMPLATE OF SELECTED RADIUS

7

ABOUT d/10

1¼ d

FIG. 10.8

THIRD ANGLE PROJECTION

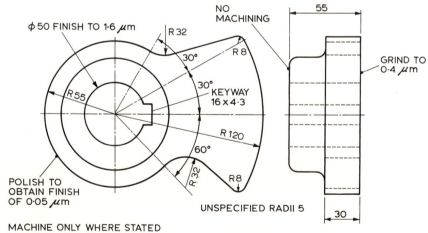

φ 50 FINISH TO 1·6 μm

R 32

NO MACHINING

55

R 8

30°

30°

KEYWAY 16 x 4·3

R 55

GRIND TO 0·4 μm

R 120

60°

R 32

R 8

POLISH TO OBTAIN FINISH OF 0·05 μm

UNSPECIFIED RADII 5

30

MACHINE ONLY WHERE STATED
ALL DIMENSIONS ARE IN MM
SCALE 1/3

FIG. 10.11

ISO Metric hexagonal nuts, bolts and screws

Dimensions in millimetres

Thread dia	Basic width across flats 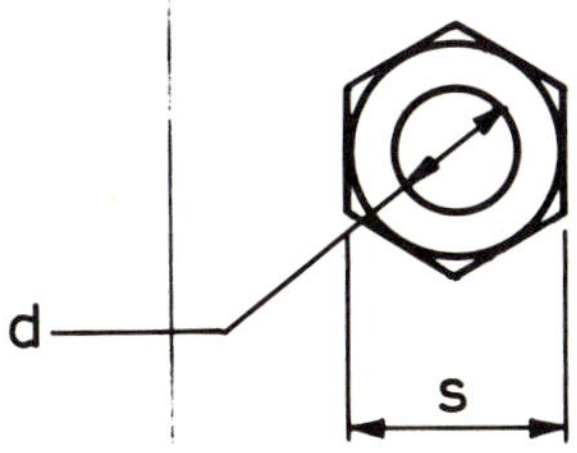		Height of head 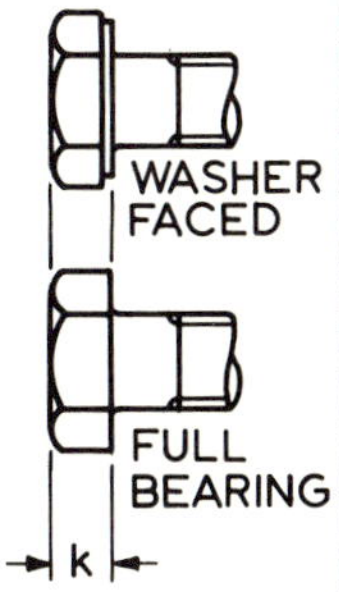	Thickness of nut 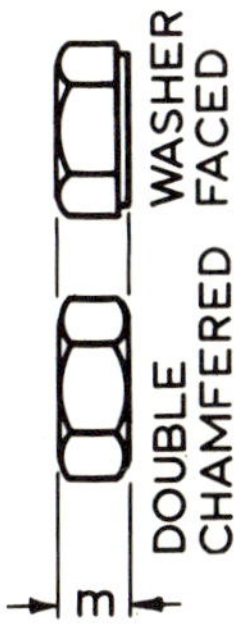	Designation
	Normal	Small			(see Fig.10.7 for pitches)
3	5·5		2	2·4	M3
4	7		2·8	3·2	M4
5	8		3·5	4	M5
6	10		4	5	M6
7	11		5	5·5	M7
8	13	12	5·5	6·5	M8
10	17	14	7	8	M10
12	19	17	8	10	M12
14	22	19	9	11	M14
16	24	22	10	13	M16
18	27	24	12	15	M18
20	30	27	13	16	M20
22	32	30	14	18	M22
24	36	32	15	19	M24
27	41	36	17	22	M27
30	46	41	18	24	M30
33	50	46	21	26	M33
36	55	50	23	29	M36
39	60	55	25	31	M39

FIG. 10.9

Machining and surface texture symbols

Machining symbol Used to indicate that a surface is to be machined, without defining either the surface texture grade or the process to be used. The symbol is applied normal to the surface or a projection line from the surface.	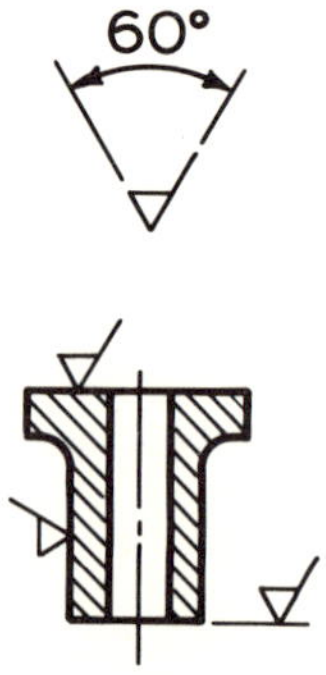 60°
General instructions Used when most, or all surfaces are to be machined.	ALL OVER EXCEPT AS STATED
Surface texture symbol Used to indicate (in micrometres) the maximum acceptable roughness of the surfaces that are to be machined.	0·8
Particular process requirement symbol Used when no other manufacturing process is to be used.	0·05 LAP
Optional process symbol Used when a particular surface texture is required but machining is not essential.	6·4
Prohibition of machining symbol Used when a particular surface texture is required and the surface must not be machined.	3·2 DO NOT MACHINE

FIG. 10.10

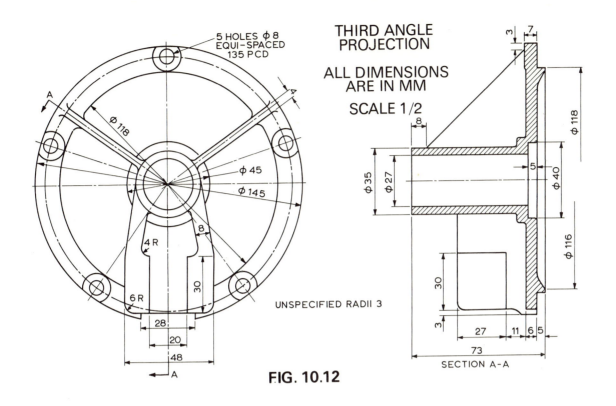

5 HOLES φ8
EQUI-SPACED
135 P C D

THIRD ANGLE
PROJECTION

ALL DIMENSIONS
ARE IN MM

SCALE 1/2

φ 118

φ 45

φ 145

4 R

8

30

6 R

28

20

48

UNSPECIFIED RADII 3

A

A

3

7

8

φ 118

5

φ 35

φ 27

φ 40

φ 116

30

3

27

11

6 5

73

SECTION A-A

FIG. 10.12

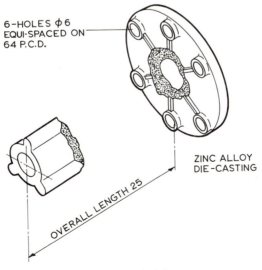

6-HOLES φ6
EQUI-SPACED ON
64 P.C.D.

ZINC ALLOY
DIE-CASTING

OVERALL LENGTH 25

FIG. 10.13

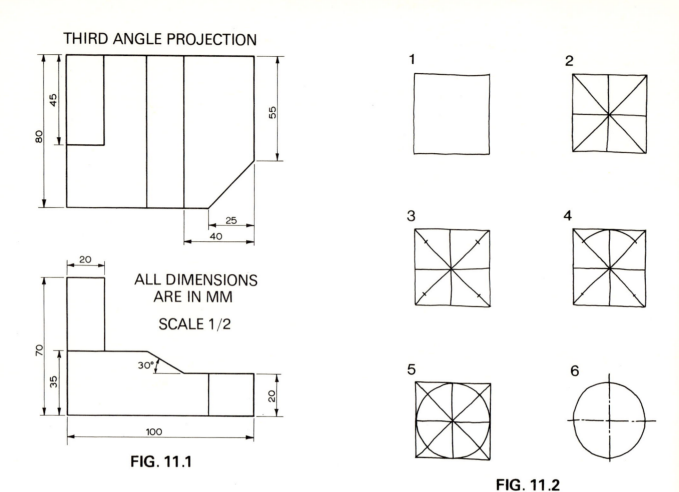

THIRD ANGLE PROJECTION

ALL DIMENSIONS
ARE IN MM

SCALE 1/2

FIG. 11.1

1

2

3

4

5

6

FIG. 11.2

ALL DIMENSIONS ARE IN MM

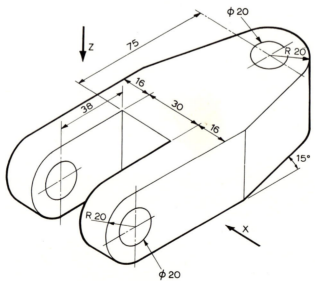

FIG. 11.3

CHAPTER 11

Sketching

At the design and development stage of any engineering product, a great deal of communication takes place by means of sketches: both freehand and made with the aid of a straight edge. The sketches may be of views in orthographic projection, or pictorial drawings made by one of the methods treated in Chapter 8. Graph paper and, for example, isometric ruled paper are very useful aids to sketching but they tend to make the sketch less clear and are not always readily available. Students are advised to practice sketching on plain paper and to use an H or HB pencil with a long conical point. Engineering sketches do not require artistic talent; they do require a methodical approach and quite a lot of practice.

When sketching vertical lines the forearm and elbow should rest on the drawing surface; for horizontal lines the most comfortable action is a sliding action of the forearm, lightly sketching segments about 50 mm long at a time. Final lines are strengthened by applying more pressure. Most people find that horizontal lines are the easiest to draw so, whenever possible, the sketching sheet should be rotated to convert any sloping lines, and especially one downwards and to the right, into a horizontal line. Orthographic views should be lined in by first drawing the rectangles that will fit around the views. Once two or more rectangles are in projection, it is not so difficult to maintain proportions and alignment.

EXERCISE 11.1

Sketch the two views shown in *Figure 11.1*. Make the views about half full size: it is more important to maintain the correct proportions than to have a precise scale.

Sketching circles

Circles may be sketched using the steps shown in *Figure 11.2*. Step 1 is to sketch a faint square, of side equal to the diameter of the circle. Step 2 is to draw in the bisectors and the diagonals of the square. Step 3 is to locate points on the diagonals at the correct radius from the centre: this can be done by eye or by using a scrap piece of paper if a particularly neat circle is required. Step 4 is to sketch a thin arc through about three adjacent points. Step 5 is progressively to rotate the paper to a comfortable position and to continue the arc. Step 6 is to strengthen the arc carefully to the required thickness of line and to erase any unwanted construction lines.

EXERCISE 11.2

Sketch views of the object shown in *Figure 11.3*, as seen in the directions of arrows X and Z. Use any suitable scale but try to keep to the proportions of the figure.

THIRD ANGLE PROJECTION   ALL DIMENSIONS
ARE IN MM

SCALE 1/2

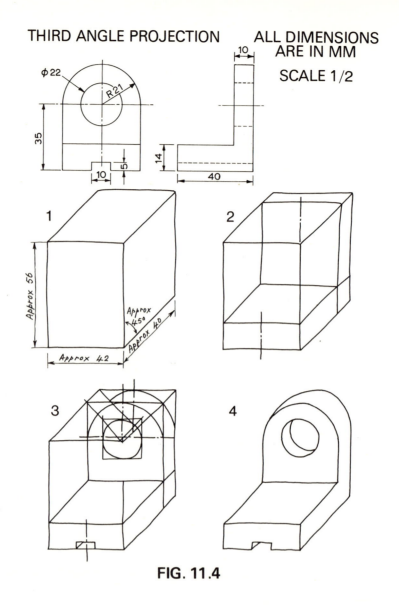

FIG. 11.4

THIRD ANGLE PROJECTION

ALL DIMENSIONS ARE IN MM

SCALE 1/2

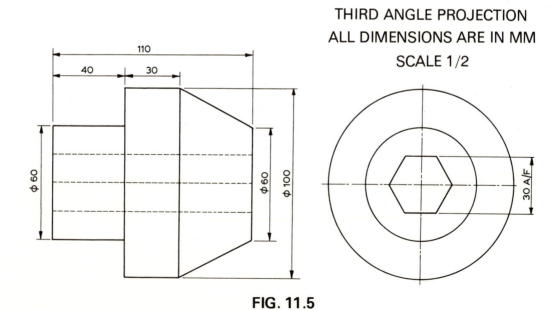

FIG. 11.5

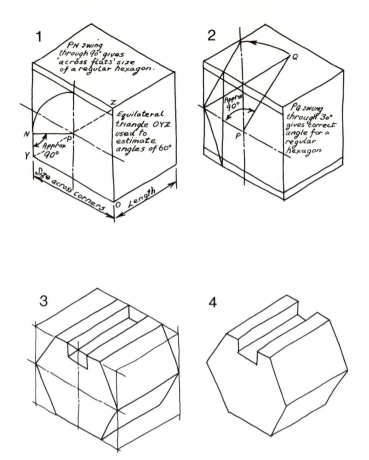

FIG. 11.6

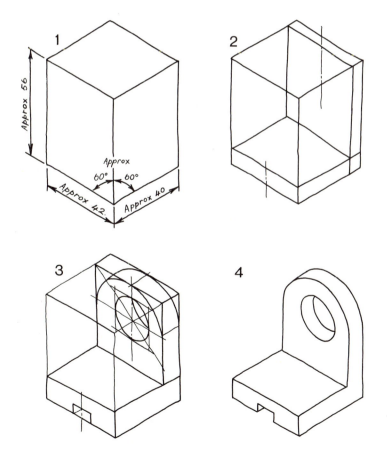

FIG. 11.8

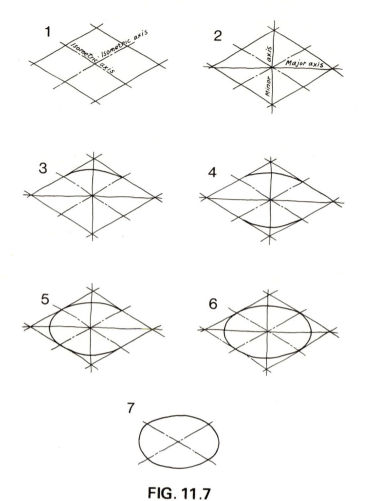

FIG. 11.7

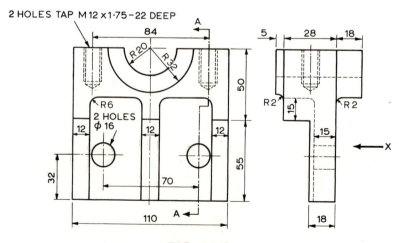

FIG. 11.9

Pictorial drawings – oblique sketches

A sketch of this type shows the front face and surfaces parallel to it in their true shape (see Chapter 8) and is particularly suitable for objects having circular or other curved features in these planes. The general method for sketching pictorial drawings is first to sketch a rectangular block of material of length, breadth and height equal to those of the object. The details can then be located on the appropriate planes and the block is, in effect, cut away until the desired result is achieved.

Figure 11.4 shows a detail drawing of a component and the various stages in the production of a "Cavalier" oblique sketch.

EXERCISE 11.3

Make an oblique sketch of the cone-clutch component shown in *Figure 11.5*. The hexagonal hole should be sketched by first drawing a circle of diameter equal to the size across flats; the side of the hexagon is then constructed equal to the radius of the circle.

Pictorial drawings – isometric sketches

An isometric sketch begins with the three isometric axes, one sketched vertically and the others at 60° from it. As with oblique sketches, the first stage is to sketch a rectangular block of material of the appropriate proportions. Thereafter, any detail is inserted by locating the required isometric plane and placing the particular lines, polygon or curve in this plane. *Figure 11.6* shows the stages in the sketching of an isometric drawing of a slotted regular hexagonal prism.

Circles in isometric planes are sketched by the method shown in *Figure 11.7*. Step 1 is to enclose the required circle with an isometric square (a rhombus) with sides approximately equal to the diameter of the circle. Step 2 is to bisect the isometric axes in order to locate the major and minor axes of the ellipse. Step 3 is to draw a curve, symmetrical about the minor axis and tangential to the isometric square. Step 4 is to mirror this curve about the major axis. Step 5 is to draw a third curve, symmetrical about the major axis and tangential to the isometric square. Step 6 is to mirror the third curve about the minor axis. In step 7, the unwanted construction lines are removed and the ellipse is boldly sketched in.

Figure 11.8 shows the stages in the production of an isometric sketch of the same object for which an oblique sketch was made in Fig.11.4

EXERCISE 11.4

Make an isometric sketch of the cone-clutch component shown in Fig.11.5. Refer to Fig.11.6 for the construction of the hexagon.

EXERCISE 11.5

Sketch, in first angle projection, views of the component shown in *Figure 11.9*, as seen in the direction of arrow X and on section plane A-A.

ALL DIMENSIONS ARE IN MM

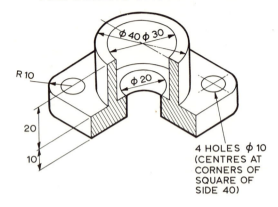

R 10

φ 40 φ 30

φ 20

20

10

4 HOLES φ 10
(CENTRES AT
CORNERS OF
SQUARE OF
SIDE 40)

FIG. 11.10

ALL DIMENSIONS
ARE IN MM

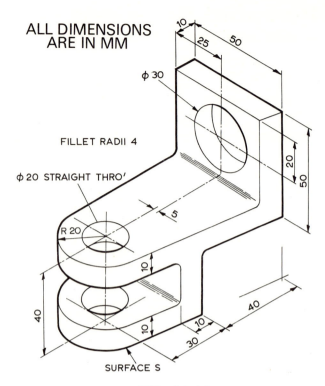

10

25

50

φ 30

FILLET RADII 4

φ 20 STRAIGHT THRO'

R 20

5

20

50

10

40

10

40

10

30

SURFACE S

FIG. 11.11

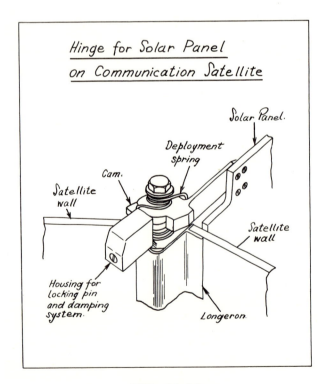

Hinge for Solar Panel
on Communication Satellite

Solar Panel.

Deployment
spring

Cam.

Satellite
wall

Satellite
wall

Housing for
locking pin
and damping
system.

Longeron.

FIG. 11.12

EXERCISE 11.6

Make an oblique sketch of the sectioned component shown in *Figure 11.10*.

EXERCISE 11.7

Sketch a new isometric drawing of the bracket shown in *Figure 11.11* so as to reveal the underneath surface S.

The sketching of ideas

The most useful function of sketching is the communication of an idea for the solution of a particular problem. Many students are reluctant to make sketches because they have great difficulty with the double task of working out the idea and putting it on paper in a reasonably presentable form. This is a skill that will only be acquired after plenty of practice but those who already possess this skill are too often unwilling to admit that they ever make intermediate sketches. *Figure 11.12* shows the sort of superb sketch often displayed with a mixture of pride and surprise that the student cannot achieve the same standard. Many people tend to develop the idea in a series of sketches (which are later destroyed) like those shown in *Figure 11.13*. Students are advised to make intermediate sketches and to be most sceptical of anyone who implies that they are unnecessary.

Intermediate sketches also help the designer to clarify the problem in his own mind. Very few people can plan, visualize and then sketch a solution to a problem without first producing an inferior sketch which triggers a new idea and enables a better sketch of an improved design to be made.

EXERCISE 11.8

Figure 11.14 shows a 13 amp plug (fused for 5 amps) with the cover removed. The normal method of retaining the cable in such a plug is illustrated: a fibre strip is cramped down on to the cable by means of two self-tapping screws. The function of the cable-retaining device is to prevent the transmission of tension in the cable to the electrical connections.

Devise and sketch an improved system of cable retention.

EXERCISE 11.9

Figure 11.15 shows two views of a magnetic tape spool. A storage rack is required for twelve such spools: each spool is to have its own compartment. When the spool, wound with tape, is inserted into the compartment the end of the tape is to be gently held so that it cannot come unwound when the rack is being transported. The tape will be wound to a radius of between 55 and 62 mm.

Make sketches of one compartment and its tape-holding system.

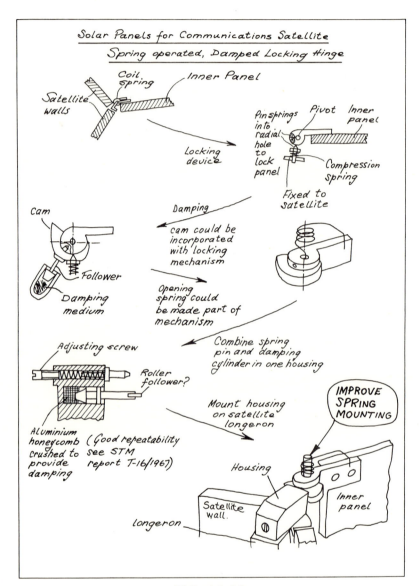

FIG. 11.13

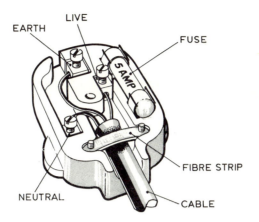

FIG. 11.14

THIRD ANGLE PROJECTION

ALL DIMENSIONS ARE IN MM

SCALE 1/2

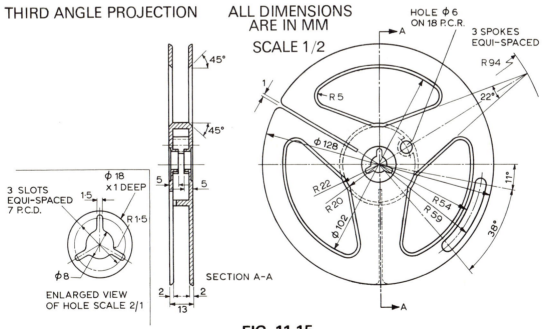

FIG. 11.15

CHAPTER 12

Introduction to Synthesis

The earlier chapters of this book are concerned with the engineering drawings and sketches that, apart from words, are an engineering designer's main method of communication. In the following chapters a simple design project will be examined from its beginnings to the detail drawing stage.

Design is a difficult word to define and means different things to different people. For these reasons, the term "synthesis" has been introduced to emphasize that the process to be examined is the very opposite of analysis. Synthesis will be taken to mean the building up of a concept of a piece of hardware which will economically fulfil a well-defined function. Synthesis is a truly creative activity in which something that meets certain requirements is gradually evolved. It is this creativity that gives so much satisfaction to the successful designer.

It is possible to identify various phases of a design project but they cannot be clearly defined and, in practice, they overlap and interact to a great extent. However, it is convenient to treat these phases in separate chapters, drawing attention to the areas where they are related. The phases are:

1. Discovering the true nature of the problem posed.
2. Determining whether the task is one that the organization will be able to carry out.
3. Examining the factors that will influence the design.
4. Finding possible solutions and selecting the most suitable one.
5. Preparing an overall design and dividing up the work that it will entail.
6. Making decisions on items of increasing detail, in consultation with the other people affected.
7. Directing the manufacture of the required quantity.

In the following chapters these phases are illustrated by looking at the design of an automatic toothbrush: a surprising example, at first sight. However, it has four advantages which, in the author's view, outweigh any objections that it is not a serious engineering project.

1. Virtually everyone is familiar with the operation that the device is to perform.
2. It is simple enough to treat fully in the available space and yet sufficiently complex to illustrate many important points.
3. As the parts are normally lightly loaded, the student does not require a knowledge of anything more than elementary dynamics and stress analysis and the project is suitable for the early stages of a course.
4. A teaching institution could purchase some example(s) of such a product and generate added interest in the project.

The exercises at the ends of the following chapters form a continuing development of the project. The dialogue passages may be read or recorded for playback to students.

CHAPTER 13

Problem Definition

The first phase in any design project is the process of discovering the true nature of the problem posed. Problems originate either inside or outside the organization which is to undertake the design: they come from a "customer" or from another department within the organization. For convenience, the originator of the design project will be referred to as the "customer", regardless of their relationship to the organization that is to carry out the design.

One of the major factors influencing the way in which a design problem is tackled is the way in which the problem is formulated. Many wrong decisions have been made and a considerable amount of time has been wasted by a poorly defined or downright misleading statement of the problem. A classic example is the "gas cylinder" story. The oxygen used in a small factory was delivered in liquid form in the usual heavy steel cylinders. Due to a reorganization, the process for which the oxygen was required was transferred from the ground floor to the second floor of the building. At first the cylinders were raised, by an improvised lifting tackle, up the outside of the building and into a second floor window. Following an accident, it was found that the cost of installing a proper lift was prohibitive and it was decided to design a "walking-upstairs" carriage for the cylinders.

Fortunately, at this stage someone asked the question: "Is it liquid or gaseous oxygen that we require on the second floor?" The answer was "Gaseous oxygen", and to design and install suitable valves and piping was a cheap and straightforward exercise.

In the above example, the error in the original definition of the problem is fairly obvious and was easily uncovered; unfortunately this is not always the case the the incorrect definition goes unchallenged. It is extremely important to start a design with the true needs of the situation established. No amount of ingenuity or technical skill can remedy the situation if the design starts off on the wrong footing.

The situation is often complicated by the fact that the ultimate user of the design either does not know his exact requirements or has difficulty in communicating them to the designer. For example, the average motorist can usually describe his requirements in only a negative fashion - the features of a particular vehicle that he does not like. His communication with the designers is through the indirect techniques of market research and the uncertainties of the mass media. Admittedly, this example is rather special but the situation is not necessarily better when the design is tailored for a particular customer. A company that sells internal telephone systems has frequently found that its customers have a very poor idea of which of their employees need to be able to contact others.

Therefore, the first step in the design process is to ascertain, so far as is economically possible, the true needs of the particular situation.

PHASE 1 Discovering the True Nature of the Problem Posed

The following discussion illustrates how one organization might tackle phase 1 of the automatic toothbrush project.

Make notes of the important points that emerge from the meeting. Then, read through the comments printed in italics.

Compare your notes with those produced by the "Director of New Projects" (*Figure 13.1*).

(To aid identification, the people present at meetings concerned with this project use each other's first name; the attendance lists are always arranged in alphabetical order.)

3/12/71

Meeting to discuss enquiry from Halworths Ltd.

Contact: Mr. V. Cabot — Director of marketing
Wants to expand in toiletries.

Proposed product: Simple Automatic toothbrush.
Interchangeable brushes.

Selling points: Appeal to children, but not a toy.
Low price — well under £3.

Quantity: 50 000/year ?

Delivery to commence: Sept '72

Priority: No special terms to secure order.

Next meeting: 11.30 — 10/12/71

FIG. 13.1

Notes made by the 'Director of new projects'

Scene: An engineering company's conference room

Present:
Andrew - Director of new projects
Brian   - Market research
Charles - Industrial design
Don   - Mechanical engineering
Eric   - Electrical engineering
Fred   - Production engineering

Notice which specialists have been invited to the meeting. As will emerge, even at such an early stage they all have something to contribute.

ANDREW. Now you're all here, let's make a start. The purpose of this meeting is to discuss an enquiry we've have from Halworths about an automatic toothbrush. I think they'll ask us to tender and it would be a very useful order if we got it. Brian and I went to see their Marketing Director to find out what they've got in mind so I'll ask him to give you the details - Brian?

Andrew deliberately refers to the project as "an automatic toothbrush". The adjective "electric" would channel everyone's thoughts towards one particular solution which the customer has not specifically requested. Halworths may be content with any satisfactory device that replaces the manual brushing action.

BRIAN. Just a few words about Halworths first: as you know they have retail branches in most large towns and they sell most things for the home, the main exceptions being furniture and the more expensive electrical goods. They sell a lot of things with their own brand name "Magpete". Their profits have been a bit static of late and so they want to expand the cosmetic and toilet preparations side of their operations.

It is important that the designer has not only general but also special, up-to-date information about the customer.

FRED. Hence the toothbrush?

BRIAN. Right. Now their Marketing Director is a chap called Cabot - he's new and he's very energetic. He's had a survey done and he reckons that the automatic brushes on sale at the moment are too expensive and very poorly advertised.

CHARLES. What does he call "expensive"?

BRIAN. The cheapest is about three pounds but you can pay over twelve pounds.

ERIC. Is that for one with a battery recharging unit?

BRIAN. That's right. There are about ten models on the market - they all run off batteries but two of them have the rechargeable type.

ANDREW. That's really why I invited you Eric - there wouldn't be much electrical engineering involved in any other type.

ERIC. Well, if your Mr. Cabot wants to sell at a really low price, I think you can forget rechargeable batteries: the nickel-cadmium cells alone would push the cost well up.

ANDREW. No, I want us to keep open minds at this stage. I'm not even convinced that the brush should be battery powered; but we'll get to that later.

There is always a tendency to start thinking about possible solutions to the problem. To a certain extent this is necessary but it can tend to limit creative thought which is why Andrew has decided to discourage it.

BRIAN. Cabot reckons, and I've checked on this, that all the research done by dental health experts leads to the conclusion that automatic toothbrushes are no better for your teeth and gums than ordinary brushes, so he wouldn't be able to plug the improved health angle in his advertising. However, kids love 'em, they're fun to use, and they certainly don't harm teeth. So what Cabot wants is something mainly for the child under twelve, at a low price, but better than a toy so that the Mums and Dads will use it too.

CHARLES. Thank goodness for that! I shan't have to start designing Mickey Mouse battery holders.

ANDREW. I thought Dougal and Zebedee were the current favourites! Please can we postpone a decision on the power supply though. Carry on Brian.

BRIAN. Well, I got my staff to look at sales of automatic toothbrushes and I'm now inclined to agree with Cabot. When you look at British homes, even those with comparatively modest incomes, they've often got televisions, transistor radios, fridges and even electric razors, but less than one per cent have automatic toothbrushes. People either aren't aware that they exist, or think they're very expensive, or can't see any particular advantage in buying one.

A significant number of homes where there is an electric toothbrush got them as Christmas presents.

It is risky to believe everything that the customer says; some of it may be pure speculation and so Brian has used his resources to make a check. There is, of course, an economic limit to how far this may be done.

DON. How does that help us?

BRIAN. The point is that, at the moment, most people don't actively want to own an automatic toothbrush, but I think that Halworths could interest children in them and their parents would buy them if the price is low enough and they can use it too.

CHARLES. So we'd have to produce something that appeals to children, but not exclusively so, and with interchangeable brushes so the whole family can have a go?

BRIAN. Exactly, and the selling price must be considerably less than three pounds.

FRED. Plastic mouldings - a metal case is right out. What quantity would Cabot want?

ANDREW. He hedged, but I think we should think in terms of about fifty thousand a year.

FRED. I see - that's fair enough. Now, that's a production rate of about two hundred a day - quite a lot of our capacity. How soon does he want to start selling them?

A design of which only one will be manu- factured poses a very different problem from one which will be made in large quantities. Fred's question is far from selfish.

ANDREW. I got the impression that, other things being equal, the order will go to the firm that can let him have a reasonable supply to start test marketing next Autumn - about nine months, in other words.

Similarly, although there is clearly a minimum time required to produce any- thing at all to solve the problem, the longer the time available the greater the number of alternative designs that can be considered.

BRIAN. I told you he was energetic.

DON. Looks as if we shall have to be too. How badly do we need this contract, Andrew?

ANDREW. Well, we only want it if the profit margin is right. Cabot wants something really cheap, but we want our normal profit on the deal - if not, then there are plenty of other new pro- jects in the pipeline. Any other ques- tions? No? Now, I'd like us to meet here again at the same time next week. Don, Eric and Fred, will you let me have your comments on the project then. I'd also like ideas from you Charles about how we make it appeal to children. Meeting closed.

Notice that the nature of the problem posed depends on the organization under- taking the design as well as on the customer. This company can easily afford to lose the order; to a company that could not the problem would be very different.

The attendance at this and subse- quent meetings is not intended to ill- ustrate good or even typical design pro- ject organization. A particular "char- acter" is present if it would be poss- ible (not necessarily desirable) for him to work on the project at that stage.

CHAPTER 14

Problem Acceptability

The second phase of a design project is a process of determining whether the task is one that the organization concerned will be able to carry out. This will depend upon the answers to several questions, for example:

What will we be expected to produce?

Is the project commercially attractive to the company?

Are there any special advantages to be gained by undertaking this project?

Do we have or can we acquire the necessary technical expertise for the design?

Do we have or can we acquire the necessary equipment for the development and manufacture?

Do we have or can we acquire the necessary skilled manufacturing personnel?

Do we have sufficient time in which to produce what is required from us?

Some of these questions will be very difficult to answer until some sort of solution to the problem has been examined. Unfortunately, the process of finding any solution will cost money and there is a limit to what can be spent. If the project is very large the customer often agrees to pay for the preliminary work involved: a design or feasibility study. With smaller projects the initial cost may have to be borne by the company, and managerial skill is needed in order to decide what should be spent so as to have a chance of winning a profitable order.

At this stage the engineering departments of the organization can give valuable advice to the management, provided the latter supplies sufficient, up-to-date information. Often, the customer's ideas and requirements are gradually evolved during this period and the technological staff need to know the relevant changes that have occurred.

PHASE 2 Determining whether the task is one that the organization will be able to carry out

This exercise follows on from the Phase 1 exercise; the same organization has now reached Phase 2 of the automatic toothbrush project. The Director of New Projects has arranged a meeting.

Make notes of the important points that emerge from the discussion.

Read through the comments that appear in italic.

Compare your notes with those produced by the Director of New Projects (*Figure 14.1*).

Scene: An engineering company's conference room

Present:
Andrew - Director of new projects
Brian - Market research
Charles - Industrial design
Don - Mechanical engineering
Eric - Electrical engineering
Fred - Production engineering

This meeting was arranged at the end of the Phase 1 meeting.

ANDREW. The first thing I want to say is that we have received an invitation to tender for the Halworth's Automatic Toothbrush. I don't like abbreviations as a rule but I can't keep saying Automatic Toothbrush, so I suggest we refer to this project as the ATB. Now

Halworths want us to quote for a test marketing batch of ten thousand: the cost of basic tooling will have to be recovered with that number. They also want figures for possible future production at the rate of fifty thousand and one hundred thousand a year. Some of

10/12/71

<u>Meeting to discuss feasibility of Halworth's</u>
<u>ATB project</u>

New information: Quotation invited for initial batch of 10 000 and possible 50 000 and 100 000 / year. We supply ATB and 4 brush-heads — no stand.

Mechanical problems:
1) Convert rotary motion to oscillating motion
2) Seal against water
3) Lubricated for life.
4) Fairly strong casing
5) Brush heads
Don advises go ahead.

Electrical problems: Sealing of switch (if any)
Eric says O.K.

Production problems: Initial batch ~~too~~ rather small
Fred says carry on.

Industrial Design problems:
1) Selection of colour and texture
2) Choice of shape if engineers want awkward controls.
Charles does not need consultants.

Action: 1) With Halworths
a) accept invitation to tender
b) 20 000 initial batch — how do they react?
c) Sub-contract brush-heads?

2) Internal — Arrange meeting with George, Harry etc. to discuss requirements.

FIG. 14.1

Notes made by the 'Director of new projects'

that would be for associated retailers in Germany and Holland.

This new information is part of the "problem definition" rather than the "problem acceptability" phase. The data for a design project never arrive in a neat, tidy package: the customer's requirements tend to alter and other information has to be laboriously unearthed.

FRED. Your Mr. Cabot thinks big.

ANDREW. Brian thinks the market is there too.

BRIAN. Yes - I've been having a look at some surveys made in the U.S.A. and Canada. There seems to be a kind of snowballing of demand after about five years of serious sales promotion.

CHARLES. Isn't it serious at the moment?

BRIAN. Hardly. When and where did you last see one advertised?

CHARLES. There was a display package in a Chemists I noticed last Saturday - but then I was looking for one.

BRIAN. Exactly. Even then you have to go to the larger Chemists and departmental stores.

ANDREW. Don, can I have your comments on feasibility?

DON. Well, trying to keep an open mind about the source of power, let's assume that we have to convert rotary motion to oscillating motion at the actual brush.

Don has succeeded in identifying one of the problems that will probably have to be solved regardless of any decision on the source of power. This is an important part of Phase 2; it is usually uneconomic to produce solutions to the problem and then decide if the problem is acceptable to the company.
The people at this meeting may visualize the product in the form of an energy flow chart:

BRIAN. All the models now on the market have that. The bristles oscillate in a vertical plane - when the thing's being used normally.

DON. Well, that shouldn't be too difficult to achieve. The power required is very low. The motion doesn't have to be precise. The other mechanical aspects are sealing the mechanism against water, saliva, etc., and providing sufficient lubrication for the life of the ATB. I'm assuming that the consumer can't be expected to oil the thing - or even to clean it, however infrequently?

BRIAN. I agree: the sort of customer that Cabot has in mind will either take it to pieces or almost completely neglect it.

DON. In any event, the sealing is simplified if no extra lubricant is required. Now the other mechanical problem will be the casing, and the greatest loads are almost certainly going to be due to dropping the thing on the bathroom floor. I think we could provide something strong enough to tolerate a fall of one metre on to say, a covered wooden floor; but if someone knocks the thing off a shelf that's about three metres above a stone or concrete floor - well, that'll be the end of it.

Don is speculating about the greatest loads. He has not mentioned a fall into a hard bathroom basin and no one raises this point. Children do not stand very much above such a basin and so Don's idea of the greatest load could be the logical one.

BRIAN. I think that's perfectly acceptable.

ANDREW. Don, what about the actual brushes?

DON. I'm going to assume that we will subcontract the brush heads. The spindle from the ATB will have to have a square end, or something of the sort, and the brush suppliers will have to hold the square hole at the end of the brush to certain limits, so there's a slight push fit.

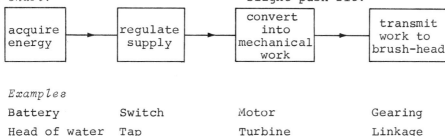

Examples

Battery	Switch	Motor	Gearing
Head of water	Tap	Turbine	Linkage
Spring	Brake	Motor	Chain

FIG. 14.2

This is a very important part of Phase 2: the company needs to recognize the parts of the project that it should not tackle. A plastic moulding containing nylon bristles is clearly a task for a specialist company.

FRED. That's easy enough with the sort of plastic mouldings that they use.

ANDREW. Right. What's your verdict then, Don?

DON. Go ahead. It's certainly something we can tackle with our present resources - except for the brushes.

ANDREW. Fine. I'll make a note to have a word with Halworths about that. Eric, can I have your comments?

ERIC. If the ATB is to be electrically powered and sell at the sort of price that Cabot has in mind then we can forget both mains operation and rechargeable batteries. For one reason, mains insulation and motors are too costly and so are nickel-cadmium batteries. For another reason, a more important one in my view, would you buy something like a child's toy which is plugged into the mains and the other end goes into the mouth?

Eric has had to come out strongly against certain possible solutions before he can advise on the acceptability of the problem. Unlike Don, the electrical engineer's tasks (see flow chart, page   ) are greatly affected by the nature of the solution and he is endeavouring to narrow the field.

ANDREW. Excellent point Eric: technically it could be quite safe but psychologically, it's all wrong. What have you got against rechargeable batteries though, apart from cost?

ERIC. We've just agreed that the ATB is going to suffer considerable neglect. My information is that you need to recharge at least every two days. The two things are incompatible.

ANDREW. Fair enough. Now, what sort of areas could that leave for electrical engineering?

ERIC. The only viable electrical power source is a subminiature motor driven by a consumable zinc-carbon battery. Now there is absolutely no point in making the motor ourselves - we could buy one from Japan or Switzerland at a fraction of the <u>cost</u> of making it here.

When Eric's arguments against certain solutions are accepted, he is able to analyse the problems that remain in much the same way that Don did. Again, there is a recognition of the work that

the company should not tackle.

ANDREW. Do you agree, Fred?

To Andrew, the possibility of his own company making the motor is still alive: it is the sort of work that he feels could be acceptable and so he takes a second opinion.

FRED. Certainly - the Japanese have swamped the market with these little motors - they <u>retail</u> at about fifteen new pence, some of them.

ERIC. That would just leave us with the battery connections, leads and switch. The only minor difficulty would be sealing the switch against water and dribble. This is certainly a job we can cope with.

ANDREW. Thank you. Fred, any problems on the production side?

FRED. Nothing too serious. I'm a little unhappy about the size of the first batch. If we ever get up to a hundred thousand a year, we could afford a lot of automation but if everything came to a halt after the first ten thousand it would mean a lot of effort and disruption for a very small return.

Fred's attitude to the problem is, necessarily, completely different. He does not need to concern himself with solutions at this stage but he is interested in the production processes and hence needs to know if the quantity can be changed.

ANDREW. Would it be better if we offered Halworths an attractive option on a second batch of ten thousand?

FRED. From my point of view, certainly. With that sort of number it's a good proposition, but it's feasible even with ten thousand.

ANDREW. Thank you. Now Charles - can you make it child-orientated?

CHARLES. A qualified "yes", I think. It's going to need quite a lot of work on this aspect. At this stage I'd be inclined to go for colours and textures that appeal to children - but not only to children. I think the idea of a special shape in the form of an animal or say, a space rocket, is out: it wouldn't be taken seriously by most adults.

Charles, too, is not greatly affected by particular solutions to the engineering problems. He has given some thought to the aesthetic qualities of all solutions.

126

BRIAN. I'd agree about the animal but I'm not so sure about the rocket, Charles; that might have possibilities.

CHARLES. Obviously, Market Research and my department will have to cooperate very closely on this point. As soon as Don's people can let me have some idea of sizes, we'll produce something for Brian to use in his surveys.

Notice the natural desire for more facts about the problem as it affects one department. Andrew discourages this line of though as it can lead to premature decisions.

DON. Can't you measure some existing products?

ANDREW. Wait please! We're moving on too fast. Charles, is this a project that we are capable of carrying out?

CHARLES. Certainly.

ANDREW. All right. Now, don't take this personally Charles: is it economically viable for us to do the industrial design or should we hand it over to someone with more experience of designing for children?

Again the problem arises: shall we do it ourselves or hire a specialist? This time a design rather than a manufacturing process is involved, but the question is just as important and needs tactful handling.

CHARLES. If we do that, I think we could end up with something that is very attractive to children but won't sell because adults would regard it as an expensive toy.

BRIAN. That's a good point, Andrew.

ANDREW. Yes, I accept that. So, it looks as if we go ahead with a competitive tender, bearing in mind the points that have emerged from this meeting. Anything we haven't dealt with?

CHARLES. The toothbrush I saw last Saturday was stored in quite an elaborate stand which also took the brush heads. Is a stand to be part of our design?

Notice how "problem definition" is still occurring, even at this point in time.

BRIAN. No. It'll keep the cost down. All Halworths want from us is the ATB and four brush heads. They will organize the packaging.

ANDREW. Right. Now our next meeting will be to discuss the requirements in detail. That's the last meeting that Brian and I will attend. I think most of you know George, one of the Assistant Chief Designers: he'll be coordinating the project from then on and Harry will be the design engineer in charge. I'll let them have notes of our meetings to date and then arrange another meeting for all eight of us that'll be - as soon as possible.

The project is now gathering momentum and more people are to be involved. They will have to be briefed about what has happened so far, and short, accurate notes are still one of the best ways of doing this.

CHAPTER 15

Design Factors

Once a decision to proceed further with the project has been taken, a third phase is in progress: a detailed examination of all the factors that will influence the design. Many companies adopt a systematic approach to this examination; some go so far as to use specially prepared forms which become part of the records of the project, and to award points to each factor according to its relative importance. Other organizations feel that they can manage just as well with less paper work, and their examination is an unstructured procedure.

There is no ideal method; what suits the particular company is right for that company. However, with the increasing complexity of human society and the products that it requires,

there has been a trend towards a more systematic approach not only to design factors but also to the whole design process.

One useful method of examining the factors that will influence the design is to use a check-list containing keywords that can be expected to trigger appropriate questions. The answers to such questions are noted down and, if desired, the factor can be rated according to the influence it will have on the design.

The answers to the questions below are translated into "requirements" during this phase of the design project. Students are advised to answer the actual questions when using the check-list for the first time; later, the factors may be expressed as "requirements".

CHECK-LIST OF DESIGN FACTORS

KEYWORDS	TYPICAL QUESTION
1 Primary functions	What is the product supposed to do?
2 Mechanical loads	To what loads will the product be subjected?
3 Ergonomics	What type of person will use the product?
4 Safety	How will the product affect the person using it?
5 Physics	What is the physical nature of the environments to which the product will be exposed?
6 Chemistry	What is the chemical nature of the environments to which the product will be exposed?
7 Geometry	Into what spaces will the product go?
8 Maintenance	What routine tasks will the user be prepared to perform in order to ensure that the product continues to function?
9 Overhauls and Repairs	How will the user react to a serious deterioration in the performance of the product?
10 Life and Reliability	For how long will the user expect the product to perform satisfactorily?
11 Running costs	How little will be user expect to pay for operating the product?

128

12 Effects on environment	What should the product not do to the environment where it is operated?
13 Appearance	To what extent is the appearance of the product important?
14 Laws, Regulations, Conditions and Standards	Has society implemented any decisions that will in any way affect this product?
15 Quantity	How many are to be produced?
16 Cost	What are we prepared to spend in order to produce the above quantity?
17 Delivery date	When is the product required?

The difference between a "design factor" and a "requirement" may be illustrated by considering the design of a filing cabinet. The question is posed: what type of person will use the product? If the answer is: "female clerical staff", then this is a design factor which may be taken into account by framing a requirement such as: "the effort needed to open and close a fully-loaded drawer should not exceed the capabilities of an average employee of this kind".

It is usually possible to divide the requirements into three main groups:

1. Vital: such requirements must be completely satisfied.
2. Important: such requirements must be satisfied but some concessions may be made.
3. Worthy of attention: such requirements may be satisfied if some advantage is gained thereby.

To continue the filing cabinet example, it could be said to be: vital to contain papers and documents, important to make it difficult to break open, worthy of attention that the eyes of the average user are 1·5 metres above the base of her footwear.

In the next chapter, the use that can be made of these groupings is examined. It is often difficult for two or more people to agree upon the importance that shall be attached to a certain requirement: they interpret the wording of the requirement in their own way. For this reason it is suggested that, although it is useful to have several minds working on the identification of the design factors, only one person should have the responsibility of converting the design factors into requirements and deciding on their importance.

The final stage of Phase 3 of the design process is that of using the requirements to write a "specification" of the product. A design specification is the means of communicating the requirements to the other people who are to be involved. The specification is usually far more technical and precise than the listing of design factors. The specification writer will decide upon target figures whenever he can, so that it will be possible to measure how well the product fulfils the specification. Considering the filing cabinet example once more, it is a requirement that the effort needed to open and close a fully loaded drawer shall not exceed the capabilities of the average female clerical employee. This could appear in the specification as: "it shall be possible to completely open or close a drawer containing 15 kg of material by exerting a horizontal force on the handle not in excess of 30 N".

The writing of a specification is a nice mixture of art and science and a detailed treatment of the subject is outside the scope of this book. A very useful booklet entitled *Guide to the Preparation of Specifications*, PD 6112, is published by the British Standards Institution.

Students are advised to try their hand at the writing of a specification by working through the check-list of design factors in the order shown. The specification can then be tidied up by identifying related requirements and grouping them together.

Phase 3 Examining the factors that will influence the design

This exercise follows on from the Phase 2 exercise; the same organization has now reached Phase 3 of the automatic toothbrush project.

1. Referring back to the Phase 1 and Phase 2 exercises, write down the design factors by working through the seventeen keywords and questions shown in the check-list.

The Director of New Projects has a meeting to examine the design factors. Compare your completed check-list with the points that emerge from the discussion and make any necessary alterations or additions.

2. Work through the check list again and, in the light of what you know about the organization working on this project, list the "requirements" that the design factors will impose and state

their importance.

From this list of requirements prepare a specification for the automatic toothbrush in a form that could be circulated to those present at the Phase 3 meeting. Try to quantify as many of the requirements as possible.

Compare your specification with that produced by Harry, the design engineer.

Scene: An engineering company's conference room

Present:
Andrew   - Director of new projects
Brian   - Market research
Charles - Industrial design
Don   - Mechanical engineering
Eric   - Electrical engineering
Fred   - Production engineering
George   - Assistant Chief Designer
Harry   - Design engineer

This meeting was arranged since the Phase 2 meeting.

ANDREW. There are just two things I want to report and then I'll hand the meeting over to George. First, Halworths have changed their minds about the initial batch; they definitely want us to quote for twenty thousand. Second, the foodmixer order from Ideal Present Club that we were hoping to get has gone elsewhere, so we need this contract slightly more than I suggested at our first meeting: we may have some extra capacity to use on it. George, the meeting is yours.

GEORGE. I think you all know the system that I like to use on these occasions. Harry will read through the titles on our check-list and also read what he's been able to get down already from Andrew's notes. I'd like you to chip in when there is something to be added. Harry?

HARRY. Thank you. One other thing, perhaps you would also tell me whether the factor is vital, or important or merely worthy of attention. First of all "Function": that's vital, by definition. I've got two things down: "Brush teeth" and "Accept other brushes". Second, "Mechanical loading": I couldn't put anything down for the power and the forces between the bristles and someone's teeth at the moment - they will have to be determined. How important is it that the brush does not break if it's dropped?

BRIAN. Worthy of attention, I'd say.

DON. I suggested a fall of one metre on to a covered wooden floor.

HARRY. Yes, I've got that. Third, "Ergonomics". "Will often be operated by children under twelve. The more difficult tasks such as changing brushes and, if battery driven, replacing batteries, will be done by adults with a wide range of dexterity". What do we think about making it difficult for a child to switch on?

ERIC. If you're thinking of a stiff switch for adult use only, we tried this once on a power tool and it's very difficult. Some great strapping eight-year-olds could switch it on and some unfortified over-forties couldn't budge the same switch.

CHARLES. I'd say make the switch quite easy to operate; after all, there isn't much harm that the brush can do.

GEORGE. It's rather messy - if you're in the line of fire.

BRIAN. You've got one?

GEORGE. Rather. They're all right once you've trained the grand-children not to switch on until it's in their mouth.

HARRY. O.K. I'll incorporate these points in the specification. Fourth, "Safety". "Casing, switch and possibly certain internal areas will be touched by user's hands. Moving brush head will go into user's mouth." Fifth, "Physical environmental conditions". "Will be used in bathrooms, washrooms et cetera".

BRIAN. Kitchens too, I should think; the sort of price that Cabot has in mind will interest people who haven't got bathrooms.

HARRY. Physically, there isn't much difference. They're all rooms with plenty of humidity and quite a range of temperature. Sixth, "Chemical environmental conditions". I've got: "Water in liquid and vapour form. Air - all types from fresh to heavily polluted".

FRED. You get all sorts of medicines and cosmetics around a bathroom.

GEORGE. We can't cater for everything. I think water is the only important one.

HARRY. Right. Seventh, "Geometry of environment". I think we can ignore that one.

ANDREW. No. I think the possibility that people might want to store them in a bathroom cabinet is worthy of attention.

HARRY. Yes, I'd overlooked that. Eighth, "Maintenance". "User will not be expected to do any". Is that "vital" or "important"?

DON. Just a moment. If we use a battery, then surely changing the battery is "maintenance"?

HARRY. Right. I'll make that: "User will not be expected to do anything more than ensure continuation of power supply".

BRIAN. I think it's rather academic whether you call that "vital" or "important". The switch could easily seize up if muck isn't removed for five years or so. We can't ensure that that won't happen.

HARRY. Point taken. I'll deal with that in the specification. Ninth, "Overhauls and repairs". What do Halworths think about this?

ANDREW. Frankly, at the sort of price we're talking about, any major repair is going to cost nearly as much as a new ATB. If we get the order, we would have a special clause in it about replacing worn or defective units. This company is not interested in doing repairs - it'll be cheaper to give dissatisfied customers a completely new unit.

HARRY. What about spare brush-heads?

ANDREW. Ah, that's something I forgot to mention. I spoke to Cabot about the fact that we would almost certainly subcontract the brush-heads. He suggested that we leave all that part of it to them; then they can deal direct with the brush manufacturers and get some packed up as spares.

GEORGE. Good! We can treat the projecting part of the spindle as an interface where our responsibility ends.

HARRY. Tenth, "Life and Reliability". We need not concern ourselves with brush-head life now. I've put down a figure of two years' continual use by a family of four. Is that too modest?

BRIAN. I think that's something to be established during the development of the product. If we have to replace too many when they are designed for two years' life, it ought to be possible to improve the life with only minor modifications.

DON. The wearing parts will be the bearings of course, parts of the mechanism, seals and the switch I suppose. If we buy a sub-miniature motor from Japan, we won't be able to increase the life of its bearings.

ERIC. There would be wear at the brushes and commutator as well but we could probably get the Japanese to produce something to our specification - if we buy in the sort of quantities that have been suggested.

GEORGE. Use your figures as a first stab then Harry, and add a note about development.

HARRY. O.K. Eleventh, "Running costs". Can anyone help here?

BRIAN. I've got some figures here on the use of batteries in some of the electric brushes that are on the market at the moment. Most of them cost about a hundred and fifty new pence a year to run.

ANDREW. How much would it cost each time the batteries are replaced?

BRIAN. Depends on the batteries; say ten new pence, on average.

ANDREW. So it would be an excellent selling point if we could cut this down, or even eliminate it altogether?

BRIAN. Certainly.

HARRY. All right, I've got that. Twelfth, "Effects on environment". I've listed "noise" and "interference with radio and T.V." as possible things to take into account.

ERIC. The latter is unlikely to be troublesome. The sort of motors we've been talking about only take a few watts - less than ten at a guess.

DON. The noise problem is closely tied up with running cost, of course. A noisy toothbrush is an inefficient toothbrush.

HARRY. Thank you - that's a point I'd overlooked. Thirteenth, "Appearance". Charles, I've got down "Design must appeal to children under twelve but must not give the impression that it is just a toy".

CHARLES. You might add that this company will want to be concerned with the shape and colour of the brush-heads as well as the hole for the spindle. We don't want an attractive design that is spoilt as soon as a brush is fixed on.

131

HARRY. Yes, got that. Fourteenth,
"Laws, Regulations, Conditions and Standards". I can't think of any laws,
regulations or conditions that are
going to affect the design - unless it's
mains operated or recharged and that
isn't feasible.

FRED. What about paint on the casing?
With children you'd have to watch the
lead content.

CHARLES. Most unlikely that we'd want
to use any.

GEORGE. It's a good point though. I
think you had better add, under "Safety"
that the casing could get bitten.

HARRY. Right. There are British Standards for batteries - it's mainly the
dimensions that would be of interest to
us. Fifteenth, "Quantity". "Initial
batch of twenty thousand. Possible future production at the rate of up to a
hundred thousand a year". Sixteenth,
"Cost". Andrew?

ANDREW. I'd like to see our unit costs
coming out at about fifty new pence.
That would mean that we'd have to produce the initial batch for ten thousand
pounds. Now that is dependent upon the
way in which the overheads are shared;
if we buy the power unit there will be
next to no overheads on that. As I said
earlier, this order is more important
to the company than appeared a few
weeks ago so I might be able to persuade
my masters to favour it.

HARRY. I'll put down fifty pence for
the time being. Seventeenth, "Delivery
Date". What can we say about that?

GEORGE. The critical factors are likely
to be the tooling up time and training
the workers on the assembly line. Fred?

FRED. That's true, but if we can have
a few prototypes by April, let's say,
there will be enough spare capacity to
start production in the Summer.

HARRY. That's a bit vague, Fred.
Could Halworths have anything at all by
next September?

FRED. Certainly - if we get prototypes
that are quite like the production
model by April.

GEORGE. Is it possible, Don?

DON. I can't answer that now; it depends too much on the power unit. If
it's a subminiature electric motor and
we only have to produce a simple mechanism and a case, then I'd say it was
definitely on. If we used - what shall
I say - clockwork, for example - then
it's going to take a lot longer.

ERIC. I'd be inclined to agree. We
could certainly produce a simple switch
and the other connections in the time
suggested, but anything more sophisticated is out of the question.

CHARLES. The same is true of the general appearance. I'd be very worried if
clockwork were used and we had to incorporate a winder and so on.

ANDREW. Let me make the position quite
clear. We won't get this order if
Halworths can't have something by next
Autumn. Therefore the design must take
that into account, otherwise there is
no point in our quoting. Nevertheless,
I'm still not convinced that a battery-
driven unit is the only way: I don't
like the word but it would be an advantage if we had a gimmick. Can you put
that point in the specification?

HARRY. All right. I won't use that
word though!

ANDREW. Good!

GEORGE. Is there anything else, gentle-
men? Good - now when you get your copy
of Harry's specification perhaps you
could let me have your comments quickly.
I want to have a "Speculation session"
to sort out possible solutions in about
two weeks' time.

SPECIFICATION

AUTOMATIC TOOTHBRUSH

Written by: H. A. Martin
Issued by : G. B. Bromley (Assistant Chief Designer)   21/12/71

Contents

1. Foreword

This specification has been written in order to set out the requirements of Halworths Ltd. for a cheap, simple automatic toothbrush that will appeal mainly to children. This document is to be used for the preparation of a tender to Halworths but is not to be a part of any future agreement with them.

2. Scope

Halworths wish to retail a package containing a driving unit and four brush-heads. This specification applies mainly to the driving unit; the brush-heads will be manufactured by a sub-contractor of Halworth's choice. We are to be responsible for the method by which the brush-heads are fixed to the driving unit and for ensuring that the design of the brush-heads is consistent with that of the driving unit.

3. Definitions

3.1 TERMINOLOGY

"adult" - anyone over the age of 12 years
"child" - anyone whose age is from 4 - 12 years
"mouth" includes lips, teeth, gums and tongue
"shall" - present tense of verb meaning "is to, unless a change is agreed with the person issuing this specification".

3.2 ABBREVIATIONS

ATB = Automatic toothbrush (driving unit + brush-head)

DU = Driving unit

3.3 MEASURING SYSTEM

Metric units are used throughout. Weights, i.e. forces due to gravity alone, are expressed in kilogrammes (kg); all other forces are expressed in newtons (N).

4. Related documents

This specification makes reference to the following documents:

4.1 Drawing of Halworth's "Magpete" trade mark (Filed as D301 292)
4.2 British Standard 397

5. Conditions

5.1 The DU will be operated in air under the following conditions:
Atmospheric pressure 99 - 105 kN/m^2
                      (990 - 1050 millibars)
Temperature   0 - 40°C
Relative humidity   20 - 100%
Domestic dust and grit in suspension.

5.2 The DU will be splashed with all types of water/saliva/toothpaste mixtures at temperatures from 0 - 100°C.

5.3 The ATB will be stored under the conditions described in 5.1 and will sometimes be in a bathroom cabinet or medicine cupboard.

5.4 The ATB will be used by children and adults. The source of power will be one that could reasonably be made available in homes where the total income is about the national average.

5.5 The only routine maintenance that will be done is to ensure the continuation of the power supply. It will be carried out by adults.

6. Characteristics

6.1 The brush-head shall be mounted in such a way that it is convenient for all users to apply it to the mouth.

6.2 The tooth-cleaning action shall, within the limits specified in 7.1, give the best results when using nylon bristles of between 9 mm and 11 mm in length.

6.3 Whatever part of the ATB is held in the hand shall have a circumference not in excess of 120 mm at a convenient gripping position, and a weight not in excess of 200 g. It shall be convenient

133

for both children and adults to grip and to operate any controls.

6.4 All parts of the ATB that can come into contact with hands, body, face or mouth shall have no sharp edges, contain no toxic materials, and be liable to cause no other type of injury to the user.

6.5 The brush-head shall fit firmly on to the DU but shall be easy to change (see 7.6).

6.6 The ATB shall appeal to children but shall not give the impression that it is a toy. The brush-heads and DU shall be visually integrated.

6.7 Halworth's trade-mark, "Magpete" (see 4.1) shall appear boldly on the driving unit together with simple operating symbols and maintenance instructions in small, legible characters. Necessary additional instructions shall appear on any internal areas of the DU to which the user is expected to have access.

6.8 The DU shall be simple and cheap to produce. (The company may limit costs to £10 000 for an initial batch of 20 000.) The ATB shall, if possible, have some unusual feature(s) that will tend to increase retail sales.

7. Performance
7.1 When the ATB is operated normally, each bristle in the brush-head shall oscillate in a vertical plane and the total vertical movement of the tip of an undeformed bristle shall be not less than 6 mm nor more than 8 mm: the locus need not be a straight line. When the bristles are not actually brushing, the frequency of the oscillations shall be between 20 and 30 Hz. At such frequencies the power available for brushing shall be at least 2 W and shall not exceed 6 W. The brush-head shall stop when a torque of not more than 30 mN-m is applied while the rest of the DU is rigidly held.

7.2 90% of production driving units shall survive a fall of 1 m on to a covered wooden floor.

7.3 The running costs of the ATB (in addition to new brush-heads) shall not exceed £1.50 per year at 1971 prices.

7.4 The free-running DU shall not interfere with a radio or television situated in an adjacent room.

7.5 The noise produced by a free-running DU shall be acceptable, in both volume and quality, to a panel of users to be agreed with Halworths and/or their representatives in this matter.

7.6 The force required to fit or remove the brush-head shall not exceed 10 N in the appropriate direction.

8. Life and Reliability
8.1 90% of production driving units shall achieve a useful life of two years when used for eight 5-minute periods each day.

8.2 The DU shall be easily modified so as to improve on the figures given in 8.1.

9. Information and After-sales service
9.1 Halworths shall be supplied with instructions for operating and maintaining the ATB, in a form that could easily be incorporated in a leaflet for the retail package.

9.2 The instructions referred to in 9.1 shall also include clauses dealing with the non-availability of a repair service and the procedure to be adopted for complaints.

9.3 Any items that the user will require to maintain the ATB shall, whenever possible, be referred to by the designation recommended in the appropriate British Standard (e.g. batteries, see 4.2).

CHAPTER 16

Problem Solving

The fourth phase of the design process is that of finding possible solutions and selecting the most suitable one. The requirements that were discovered during Phase 3 may be classed as "vital", "important" or "worthy of attention". A "possible solution" to the design problem is any solution that could satisfy all the vital requirements: no other solutions are worthy of further consideration. In practice, it is unusual for such solutions to be proposed; the term "possible solution" has been introduced as a more convenient way of defining what is a "vital" requirement. To take an extreme example, a hydrogen balloon is not a possible solution to the problem of getting a man to the moon because, by being unable to reach the escape velocity, it does not satisfy one of the vital requirements.

Therefore, at first, the search for possible solutions is focussed on the "vital" requirements. When these have been satisfied, the designers can turn their attention to the "important" requirements. To continue with the moon journey example, an (ironic) important requirement is that the cost shall not be astronomical. Such a requirement might rule out the possibility of using atomic power.

All "possible" solutions will satisfy all the important requirements but the extent to which they do so will vary. By intuition, experience or some system of points, someone has to decide upon which of the possible solutions is the most suitable. If that cannot be done, it is then necessary to look at those requirements that are worthy of attention. The most suitable solution is always a compromise between conflicting requirements and the choice is rarely obvious.

The brain can perform the most amazing tasks of keeping all the requirements in store, assessing their importance and hence judging the solutions presented to it - but it is a <u>human</u> brain; it is apt to be prejudiced on certain questions. A systematic approach to the selection of the most suitable solution can help to counteract this prejudice.

Possible solutions are found by a variety of methods but it is hard to better the human brain as a store, or more accurately, a factory for such solutions. An individual mind often performs less well by itself and other minds can act as a stimulus which improves its performance. For these reasons, the search for possible solutions is often done by a team at a meeting whereas the selection of the most suitable solution is often better left to one individual.

The group search for possible solutions will be referred to as "Speculation". It has often been found that such meetings give the best results if:

1. All evaluation is banned,
2. The people present aim at quality of ideas, not quantity, and
3. Ideas which closely resemble a previous idea are still noted down.

Phase 4 Finding Possible Solutions and Selecting the Most Suitable One

The exercises follow on from the Phase 3 exercises; the same organization has now reached Phase 4 of the automatic toothbrush project.

1. Speculate about possible methods of powering the automatic toothbrush by considering the question: in the places where the brush is to be used, what sources of energy could be made available?

Evaluate these methods using your own knowledge and the information contained in the specification. Reject only those proposals that are obviously unsuitable; do not attempt to select one "ideal" method.

2. Consider the question: How could the power be made to produce the type of motion described in the specification, clause 7.1? Speculate about possible ways of converting and transmitting the energy that is available (see Fig.14.1).

Evaluate these methods using your own knowledge and the information contained in the specification. Reject only those proposals that are obviously unsuitable; do not attempt to select one "ideal" method.

3. The Assistant Chief Designer has a meeting to consider the problems posed above. Compare your speculation and evaluation with the points that emerge from the discussion. Read through the notes in italic.

Scene: The office of an Assistant Chief Designer

Present:
George   - Assistant Chief Designer
Harry   - Design Engineer
Ian   - Designer (Mechanical)
John   - Designer (Electrical)
Ken   - Production Engineer
Lawrence   - Industrial Designer

The new personnel were chosen and the meeting was arranged in the later stages of Phase 3.

GEORGE. I think we've now dealt with all the comments that were received about the draft specification and you've all got copies of the first issue, dated 21st December. Now, let's do our first bit of speculating: I think that the first question we must ask ourselves is - in the places where the brush is going to be used, what sources of energy could be available? Speculation only please, I don't want any evaluation.

The way in which the question is posed has a considerable influence on the answers given.

Essentially, there are only three methods of acquiring energy:

(1) create or use a chemical reaction
(2) create or use a nuclear reaction
(3) create or use a force field (gravity, magnetism or electrostatics).

Electricity may be produced by any of these methods. At the present time, the many methods of turning heat into mechanical work are relatively complicated and inefficient.

JOHN. Mains electricity and batteries.

KEN. A head of water.

IAN. Some convenient sort of hydrocarbon fuel, like a butane gas cartidge for a cigarette lighter.

HARRY. Talking of cartridges - one of those soda syphon things with carbon dioxide in it.

LAWRENCE. Can I just clear up one point? It's supposed to be an <u>automatic</u> toothbrush so does that mean that the person using it must not supply the energy in any way whatsoever?

Clarification is allowed during the speculation stage. It is really a part of the process of defining the problem.

GEORGE. I don't think so - Halworths would probably accept something that used body energy, as long as the user doesn't actually have to agitate the brush-head directly.

LAWRENCE. Then I suggest storing energy in a spring.

IAN. Do you mean a metal spring?

LAWRENCE. Yes.

IAN. I'll add a fluid spring, then.

KEN. It's a very low power that's required. I wonder if there'd be enough heat from a hand to drive it?

JOHN. I very much doubt it: you need a body surface area of . . .

GEORGE. Sorry John, but that sounds like evaluation to me. Any more speculation please?

The temptation to evaluate is always present. Although the criticism may be entirely justified, it does tend to discourage other speculation. Conversely, a chairman who prevents evaluation will tend to encourage further speculation.

LAWRENCE. As I'm safe from criticism at the moment, I suggest solar energy.

HARRY. Great! We could call it "The Sunshine Brush".

GEORGE. Any other sources of energy? No? Right, let's start some evaluation - would you like to comment Harry?

HARRY. If you look at item 6.8 in the specification - "simple and cheap to produce" - I think we must reject any source of energy that has to be converted from heat into work; however you do it, it either means a lot of moving parts or some very accurate manufacturing process, or both.

Engines that convert heat into work are rarely economic at power outputs below about 10 kW. There are, of course, special applications where no other source of power is suitable.

GEORGE. I agree, so I'll cross out "butane gas", "body heat" and "solar energy" unless someone wants to defend them. Right?

HARRY. Now, when the feasibility of this project was discussed, I gather that Eric more or less convinced people that mains electricity was unsuitable: cost and psychological reasons. Would you agree, John?

JOHN. Certainly; when I suggested it I was simply answering the question that George posed.

Generally, it is better to pose the question in a way that encourages speculation than to discourage it by stating in advance that certain solutions are unacceptable.

GEORGE. That's all right, at least we haven't overlooked it. So that leaves us with "batteries", "head of water", "CO_2 cartridge", "metal spring" and "fluid spring".

HARRY. In retrospect, I'm not at all keen on the CO_2 cartridge; there would be some very large pressures and forces involved: we'd find it hard to make it safe - at any price.

IAN. Hear, hear. If we want some sort of turbine, I think water is a much better bet.

GEORGE. I've crossed out CO_2 cartridge. Now, I think it would be convenient if we refer to the four suggestions we have left as "electric", "hydraulic" and "spring" methods. Lawrence, your boss was rather worried about incorporating a winder for any clockwork mechanism - would you agree?

Once the number of possible solutions has been reduced, it is useful to adopt titles that do not restrict thought to some particular interpretation of the proposed solution.

LAWRENCE. Not entirely, no. I gather he was thinking in terms of something that looks like a key, but it need not be. It could be a cylindrical end-cap that fits flush with the rest of the casing - quite easy to incorporate.

The person who is to be more concerned with the details of the product is often in a better position to evaluate the various proposals.

KEN. Forgive me Mr. Chairman but I think it's plain stupid to even consider clockwork. For one person to use it for five minutes, that's clause 8.1, the stored energy would have to be about one kilojoule. That'll need quite a spring which'll take either a large effort or a long time to wind up, or compress. We couldn't make it economically ourselves but a suitable unit would be quite expensive to buy. Then, looking at 8.1 again, we'd have great difficulty in keeping the water out and sufficient lubricant in for a two year life. What's more, looking at 7.2, I doubt if anything like 90% would survive a fall of one metre.

Apart from timing devices and toys, clockwork motors are used successfully in cine cameras, dry razors, etc. The advantage of a systematic approach to Phase 4 is that people are expected to explain their reasons for rejecting a particular solution and the prejudiced "it won't work" attitude can be overcome.

GEORGE. Thank you for being so frank. I think you've make some good points but they do apply mainly to traditional clockwork motors. Perhaps, with a fluid spring we could avoid these problems?

HARRY. Maybe, but of course the one thing that I'm not supposed to put in a technical specification is that Halworths won't give us the order if they don't think that we could deliver by next Autumn. I think that a novel type of spring motor, that we'd have to develop if we can't buy one, would put them right off.

In any design project it takes time and money to obtain data. It is rarely possible to completely evaluate all the proposed solutions; the very fact that more data are needed is a real argument against adopting such a solution. However, there would be no technological innovation if lack of data were the only deciding factor for rejection; the future prosperity of the organization has to be considered. In other words, there are long-term dangers involved in a policy of always selecting the most straightforward solution.

LAWRENCE. Speaking as a parent, I'd say that something spring-operated that appeals to children would look very much like a toy to me.

GEORGE. Let's "recap". In favour of an energy storing spring: it's novel, wouldn't cost anything to run, and the controls could be blended into the general shape. Against: expensive and relatively delicate, lots of moving parts to give sufficient mechanical advantage for children to put the energy in, lubrication and sealing problems, and it would look very much like a toy when we'd made it - which could well take longer than other drives. I'd say reject it. Any strong objections?

An important part of Phase 4 is the summary or "recap" procedure. Unless someone is taking notes, it is very easy to forget some of the points that have been made. A summary of the arguments for and against will ensure that nothing is overlooked.

IAN. As one of the people who suggested springs, I agree - the disadvantages are overwhelming.

LAWRENCE. Seconded.

At first, it is possible to reject some of the proposed solutions because they lack some of the qualities required. It is then necessary to look more closely at the quantities detailed in the specification. One of the most important of these is the power involved. Unlike quantities such as speed, torque, deflection under load, weight, etc., that the engineer can readily manipulate, once power has been supplied to a system, it cannot be increased. Inevitably power is lost until some fraction of it is used for the required purpose. Therefore, the power available is of great importance in the selection of a source of energy. Students would do well to familiarize themselves with the means by which such powers are estimated.

	Input power (watts)
D.C. electricity	Current (amps) × Potential (volts)
Single phase A.C. electricity	Current (amps) × Potential (volts) × Power factor (ratio)
Water supply	Head (metres) × Weight flow (newtons per second)
or	Weight of head (newtons) × Velocity (metres per second)
Heat	Rate of production of heat *minus* Rate of loss of heat due to irreversibility (joules per second)

GEORGE. Good, that leaves us with a straight choice between "electric" and "hydraulic" methods. We must find out now whether the latter could meet the specification; I don't think there's much doubt that the former could. I'm thinking about the power required.

JOHN. That's quite easy: a high power 1·5 volt battery could deliver up to four amps. That would give us six watts input power - even allowing for the low efficiency of a subminiature motor, it's plenty.

IAN. I've been scribbling rapidly ever since Ken suggested a head of water. At worst we might have a level in the

cold water tank at 2·5 metres above the cold tap: if we could get a flow of 9 litres in 30 seconds - I'm thinking about when I fill my watering can, although that's from a feeble kitchen tap - that's 0·3 kilogrammes of water per second. The potential energy of one kilogramme will be 2·5 times 9·81 joules, say twenty five, and 0·3 kilogrammes per second would give us 7·5 watts. It's too good to be true.

The water pressure is another fixed quantity, associated with the available power, that is bound to have an important influence on the design.

HARRY. How much of that can you get to the brush-head: that's the question?

GEORGE. We certainly can't reject it on the question of energy available though. What does worry me is how you'd make the thing easy to fix on to a multiplicity of bathroom taps. Taking Ian's figure of a head of 2·5 metres - that's a pressure of about a quarter of an atmosphere - not difficult to get a pipe to stay on with that, but I think we're more likely to get mains pressure.

KEN. We'd have to think in terms of some special adaptor that stays clamped to the tap and you fix the ATB to that when you want to use it. There's also the problem of getting rid of the waste water.

HARRY. The turbine doesn't have to be inside the part that's held in the hand. It could be by the tap and discharge the water straight to the plug hole - flexible drive to the brush-head.

IAN. Terrible sealing problems.

The tendency to define a problem, propose a solution and criticize the proposal - all in rapid succession appears to be a basic human characteristic. Only long schooling or a strong chairman can prevent it and hence gain the benefits of a more systematic approach.

GEORGE. Wait! Harry's idea sounded like a good bit of speculating. Let's evaluate it later. Can we have some other speculation please about tap-driven systems? We must have high efficiency.

LAWRENCE. Use coaxial tubes - an inner one to feed the turbine and an outer one for the waste water.

JOHN. I think you could get a sufficiently high efficiency from a hydroelectric system - turbine and generator fixed to the tap and a flex from them to a motor in the hand unit.

IAN. Ultimately, we don't want rotary motion: why not a jet of water giving forced vibration to a spring at the required frequency?

GEORGE. All done? Right - evaluation - Harry?

HARRY. How efficient would one of these little motors be, John?

JOHN. Could be about 50%, possibly the same when you use one as a generator.

HARRY. Multiply those two fifties by the turbine efficiency and there's not much left, I'll bet. Ian's forced vibration idea sounds quite good except that there will be quite high peak loads. We probably need the flywheel effect of a turbine to cut down the speed variation.

Another advantage of a systematic look at solutions can be that, once people are used to the procedure, they save time by not overstating their case. The hydroelectric system is rejected because of the power loss involved - there was no need to add, for example, that it would double the cost of electrical components compared with a battery-operated system.

KEN. I think the coaxial tube idea would make it very easy to flood the turbine; water is going to be slow to run away on the downstream side.

GEORGE. It seems to me that we're left with turbine at tap, turbine in hand unit and forced vibration in hand unit but we reject electric transmission. I think the best thing we can do now is compare those with battery operation, John?

JOHN. Far be it from me to run down electricity but, when I first saw the specification, it struck me that a sub-miniature motor was almost a foregone conclusion and that depressed me because we would then have a product very much like everything else that's on the market. Our product would be inferior because of its low price and that would be its only selling point because, Ian'll correct me if I'm wrong, a cheap

mechanism is going to be inefficient and the cost in batteries will be well over the £1.5 per year that's stated in clause 7.3.

A battery-driven toothbrush is a demonstrably suitable solution to the problem posed by the specification; it could satisfy all the vital and important requirements. However, some of the requirements under consideration are unusual and adversely affect the suitability of a battery-driven model.

IAN. I've done some work on that very point; I've got some sketches of possible mechanisms that should be fairly cheap but they won't be any more efficient than existing ones. I doubt if we can reduce the running costs of a battery-driven model so, as John says, there's no extra selling point.

GEORGE. This would be the right moment to look at your sketches, Ian. Apart from the forced vibration proposal we shall always have the problem of converting rotation into the motion that's required.

IAN. If you look at this sketch, ATB 1, you'll see that it's a four-bar chain, ABCD. CD is in its extreme right-hand position; as AB rotates through 180°, CD will move to the left and move the tips of the bristles up through the specified eight millimetres. I'm assuming that the roots of the bristles will lie in a plane through D: that way we'll minimize the inertia of the brush-head. CD will return to the position shown as AB goes from 180 to 360°. The whole thing could fit into the circumference that we have available for gripping. I was thinking, when making the sketch, that we'd have an electric motor with its shaft roughly at the centre of a cylindrical body . . .

GEORGE. Well, let's not get too immersed in detail. Any comments about the basic idea?

HARRY. There'll be friction at the bearings supporting shafts A and D and at the pivots B and C but I don't see how you can cut it down; unless you get the oscillating motion straight from the power supply, you've got to have at least one additional shaft with its own bearings.

IAN. I think there'll be more friction from the mechanism in this other sketch, ATB 2, although there's one less part. This is an inversion of the slider-crank chain: CD is driven down and then up by the pin at the end of AB. Of course, this is a form of quick-return mechanism and CD completes its downstroke in just over half the time it takes for the upstroke - but there's nothing against that in the specification.

HARRY. I must admit it's something I didn't think of, but I don't see that it matters very much. The bristle tip speed that's implied is already quite low compared with some models on the market.

Figure 14.2 can be used to indicate that there is another area of the toothbrush problem (converting rotary to oscillating motion) that is largely unaffected by the source of energy. By identifying such areas and considering possible solutions at an early stage, it is possible to save time at design meetings.

Sketch ATB 1 is illustrated in Figure 16.1.

A four-bar chain is any plane linkage with four pivot-points. Two of the pivot-points are fixed in space and the other two move in complete circles or circular arcs, depending on the geometry of the linkage.

Sketch ATB 2 is illustrated in Figure 16.2.

A slider-crank chain is really a special case of a four-bar chain where one of the pivot-points moves in a straight line (a circular arc of infinite radius). In the form in which it is used in an internal combustion engine, the straight line (cylinder wall) is fixed in space; an inversion of this mechanism allows the straight line to move.

There are many advantages in having the writer of the specification present at this stage and fully involved in the project. A remote figure of putative infallibility is apt to be misunderstood.

KEN. It's a better mechanism from the production point of view than ATB 1 - two fewer pivots to assemble.

LAWRENCE. For the industrial design, it might be nice to have the rotating shaft and the oscillating spindle, co-axial. Could that be done?

The industrial designer can be employed in several ways: at one extreme he is ordered to "put a pretty case around this lot"; at the other he is treated as an equal member of a team and his views are considered at all stages of the project. Not surprisingly, the second policy is nearly always the more rewarding.

KEN. No kinematic problems. In both mechanisms the cranks AB could be integral with a gear wheel that's driven by a pinion rotating about D.

HARRY. More friction though - still, it would be a useful way to get AB running at the specified speed.

GEORGE. Now, I think we've gone about as far as we usefully can for the moment. The Director of New Projects wants to see an overall design quite soon. I think we may have to submit battery and tap-driven designs to him and let him find out what Halworths would prefer. The choice can't be made on economic and technical grounds alone; there's this requirement about extra selling points. John, can you start to produce an overall design for a battery-operated brush, using Ian's mechanisms?

When there is no solution that appears to be the most suitable, the choice will depend upon the requirements that were "worthy of attention". If the customer has made some ill-defined requests about characteristics that the product should have, a few definite proposals will usually produce reactions that clarify the situation.

JOHN. Yes, I think I've got all I need.

GEORGE. Fine, and Ian - will you get together with Harry and do the same for a tap-operated system?

IAN. Turbine or forced vibrations?

GEORGE. It's up to you; I want something to show Andrew fairly soon and so I'll be chasing you for those schemes. Will you make a pictorial drawing so that Halworth's people can understand it. Just show the moving parts and anything else of real importance. I want you to agree on the same orientation for both drawings so that they're easy to compare. It would be useful if you showed the same sort of spindle for fixing the brush-head to. We'll meet again as soon as the overall design has been fixed.

There is a considerable overlap between Phase 4 and Phase 5 as they are defined in Chapter 12. For the toothbrush project, two overall designs are to be prepared <u>before</u> the most suitable solution is selected.

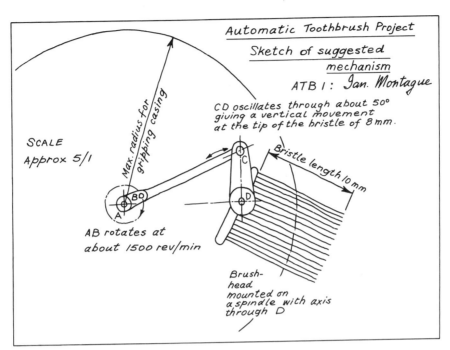

FIG. 16.1

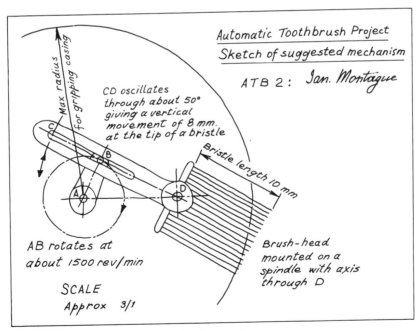

FIG. 16.2

CHAPTER 17

Solution Division

The fifth phase of synthesizing a new product is that of preparing an overall design and dividing up the work that it will entail. The overall design may be prepared in any convenient form: for a machine it is essentially a written and/or visual statement of the way in which energy is to be acquired, regulated, converted into mechanical work and transmitted to the place(s) where it is required. The overall design of a structure is a statement of the way in which loads are to be accepted and transmitted.

The purpose of the overall design is to facilitate a division of the total design work still to be done. Consider an ordinary typewriter: the overall design will show that energy is to be acquired from the typist by the act of tapping on the keys and pushing the carriage to the right. The latter energy is stored in a coil spring and regulated by the depression of a key. (The energy has already been converted into mechanical work by a biochemical process inside the typist.) Part of the energy supplied at the keys is transmitted to the type-bar as kinetic energy in order to squeeze the ribbon against the paper, another part is stored in a tension spring to return the type-bar to its original position, and another part is used to drive the ribbon along. The picture is still incomplete but it illustrates the concept of an overall design: the detail work to be done can be divided up as required.

The way in which the work is divided will depend on the specialist skills the organization has available. One common division is into electrical and mechanical areas of responsibility. The electrical work could itself be divided into power systems and logic systems. The mechanical work might be divided into structural and vibration problems, mechanisms, fluid flow, thermodynamics and heat transfer. There may be other specialist areas such as an optical system or the industrial design.

Some organizations whose products are less sophisticated prefer to divide up the design work according to the available man-hours; most of the staff have sufficient expertise to tackle any part of the design.

The physical divisions are more difficult to settle. For example, if one person is to design the carriage of the typewriter, does his responsibility end with the moving parts or should it include the guide rails on which the carriage runs?

The concept of an "interface" is very useful here. Students who are familiar with the idea of a "free body diagram" from their study of dynamics will readily appreciate that one component or subassembly can be thought of in isolation, provided the geometry, forces, torques, heat and fluid flows, etc., from and to neighbouring parts are taken into account at the "interfaces".

Figure 17.1 illustrates the interfaces between a typewriter key bar and the other parts affecting it.

Once the work has been divided, it may be necessary to establish a system for communication between the people involved, i.e. when who is to tell whom about what. Most organizations train their staff to provide each other with the necessary information but it is wise not to trust too much to people's so-called commonsense.

EXERCISES

Phase 5 Preparing an overall design and dividing up the work that it will entail

These exercises follow on from the Phase 4 exercises; the same organization has now reached Phase 5 of the automatic toothbrush project.

1.   Prepare an overall design according to the instructions given to either Ian or John at the end of the Phase 4 meeting.

Figures 17.2 and 17.3 show the overall designs that were prepared by Ian and John respectively. Compare the appropriate one with your solution.

2.   Fig.17.2 illustrates the overall design that is to be adopted for the Halworth's project. List the various types of specialist knowledge that will be required and suggest how the work might conveniently be divided. Try to define clearly where each person's responsibility begins and ends. Compare your proposals with the decisions made by the Assistant Chief Designer at the following meeting.

Scene: The office of an Assistant Chief Designer

Present: George   - Assistant Chief Designer
            Harry   - Design Engineer
            Ian   - Designer (Mechanical)
            John   - Designer (Electrical)
            Ken   - Production Engineer
            Lawrence - Industrial Designer

This meeting was arranged since the Phase 4 meeting.

GEORGE. A few pieces of information first of all - for those of you who haven't already heard on the grapevine. Ian produced this overall design, ATB 3, (Fig.17.2) which is essentially an impulse turbine driving the mechanism in ATB 1. Why did you go for the turbine rather than the forced vibrations, Ian?

IAN. Harry and I agreed that forced vibrations couldn't be very efficient without a lot of moving parts to produce a push in two alternating directions. Even if you did that, there wouldn't be enough kinetic energy available to cope with sudden peak loads.

GEORGE. Fair enough. John produced ATB 4 (Fig.17.3) which is a single carbon-zinc cell driving a subminiature motor and the output is geared down to drive the mechanism in ATB 2. Why the gearing, John?

JOHN. Mainly because there is a very cheap motor available that runs at five thousand revs per minute. The difference between its cost and the cost of a slower motor more than compensates for the extra moving part. This design could be cheap to buy but there is no other selling point.

GEORGE. And that's just what Mr. Cabot at Halworths thought, too. Andrew said that he was much more enthusiastic about driving it from the tap - but he's no fool; he pointed out all the disadvantages, especially the problem of fitting a pipe to all sorts of different taps and getting rid of the water that's used. Halworths sell a hair washing

spray at the moment and they get a lot returned by people who have trouble in fitting them to their taps. Anyway, that's the sort of design that he wants us to tender for. Thanks for your help, John, and for coming along today; I gather you weren't too keen on this design in ATB 4 and Eric's found plenty for you to do in any case?

JOHN. Oh yes, it would have been a very inferior product. I'll get back to something else. Cheerio.

GEORGE. Thanks again. Now, are there any questions before we divide up the work involved in ATB 3?

KEN. Why have you used an impulse turbine rather than some other type?

IAN. Mainly for simplicity but the specific speed is about right for the available head.

LAWRENCE. I'm rather lost, I'm afraid. This sort of arrangement is going to be quite awkward for me; do I understand that there are other possibilities?

HARRY. As I persuaded Ian to use an impulse turbine, perhaps I should explain. In this sort of turbine you let the water come out of a nozzle into air at normal atmospheric pressure. All the energy the water gained as it came down the pipe appears as velocity; there's no pressure left. The jet hits the turbine wheel and spins it round but the wheel must rotate in air, you can't have water building up and getting in the way. Now the other possibility

143

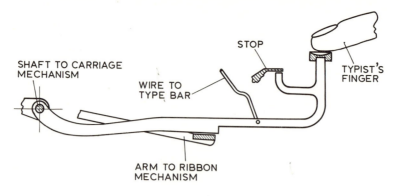

SHAFT TO CARRIAGE
MECHANISM

STOP

TYPIST'S
FINGER

WIRE TO
TYPE BAR

ARM TO RIBBON
MECHANISM

FIG. 17.1

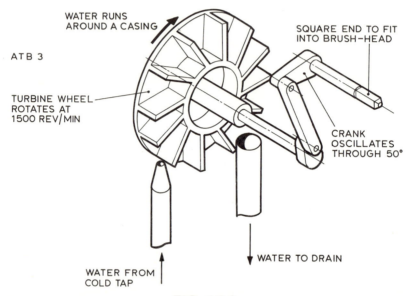

WATER RUNS
AROUND A CASING

SQUARE END TO FIT
INTO BRUSH-HEAD

ATB 3

TURBINE WHEEL
ROTATES AT
1500 REV/MIN

CRANK
OSCILLATES
THROUGH 50°

WATER FROM
COLD TAP

WATER TO DRAIN

FIG. 17.2

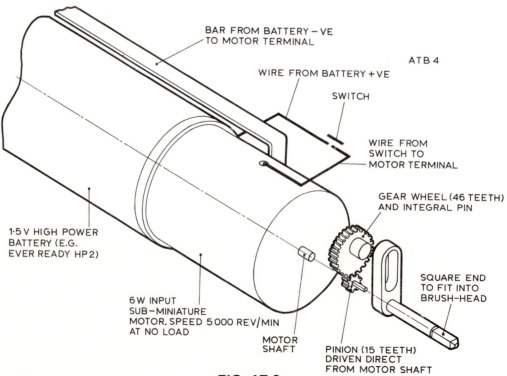

BAR FROM BATTERY − VE
TO MOTOR TERMINAL

ATB 4

WIRE FROM BATTERY + VE

SWITCH

WIRE FROM
SWITCH TO
MOTOR TERMINAL

GEAR WHEEL (46 TEETH)
AND INTEGRAL PIN

1·5 V HIGH POWER
BATTERY (E.G.
EVER READY HP 2)

SQUARE END
TO FIT INTO
BRUSH-HEAD

6 W INPUT
SUB-MINIATURE
MOTOR. SPEED 5 000 REV/MIN
AT NO LOAD

MOTOR
SHAFT

PINION (15 TEETH)
DRIVEN DIRECT
FROM MOTOR SHAFT

FIG. 17.3

is to use some of the water's pressure to drive a wheel round; we call that a reaction turbine. It's something like a propeller working in reverse: you send water over it and make it go round but you need guide vanes to make sure that the water goes in in the right direction and I think that will make the whole thing too expensive.

LAWRENCE. What is the other thing Ian said - specific speed?

IAN. It's a useful way of comparing different turbines. If you could get just one kilowatt out of a turbine from a head of exactly one metre of water, then the speed at which it was running is the specific speed. We need much less power from more than twice as much head so our turbine will be scaled down compared with one working at the specific speed. However, we know how to find the speed that a scaled-up version of our turbine would run at under the specific conditions: it works out at about twenty eight revs per minute and that helps us to choose the most suitable type of turbine. I'm afraid there's not much doubt that it should be an impulse type but don't forget that drawing ATB 3 is highly diagrammatic. For quickness the wheel has been drawn more like a paddle wheel than an impulse turbine. The water must go in in a tangential direction but it splashes about as it comes out.

KEN. I seem to remember that the little buckets that the water hits are quite complicated things. We could produce a plastic moulding at a reasonable price but you wouldn't have much mass to give the required flywheel effect.

IAN. That's what I thought; it may be necessary to incorporate a metal ring to increase the mass.

GEORGE. Can we move on then? Why this linkage, Ian?

IAN. Less friction than ATB 2 and the spindle from the brush-head ends up nearly in line with the circumference of the turbine wheel. I thought that would make Lawrence's job easier.

LAWRENCE. Yes, I think it does.

GEORGE. Good. Now the other thing that concerned me particularly was the ergonomic problem with regard to the application of the brush-head to the molars on both the right and left sides of the mouth. It's easy enough to move a battery-operated toothbrush from one side to the other but we've got problems with water inlet and outlet when we use a turbine. The inlet may get twisted up and the outlet may be held at the

top and the turbine will flood or soak the user. Anyway, Harry's done some research so I'll let him tell you about it.

HARRY. I arranged for the Methods Study people to film two children and two adults using a battery-operated brush. I'd like you all to see the film but I think I can summarize the results now as they're quite conclusive. We painted an arrow on the end of the casing furthest from the brush-head and pointing away from it, and a star on the casing at about the centre of gravity but on the area furthest away from the user's mouth. In all cases the arrow points slightly downwards for at least ninety per cent of the time and the star moves anywhere in a vertical plane within a circle of radius fifty millimetres relative to the user's mouth. So, if we had water going in at the star and coming out parallel to the arrow, it would be quite easy to handle.

LAWRENCE. Does the arrow point upwards when the user is transferring the brush from the left side to the right side?

HARRY. Never. Even the younger child - about five he was - didn't attempt to transfer like that. It's most awkward: you'd have to raise your elbow so that it was almost level with your mouth and change hands. The arrow points slightly upwards for a little while when adults are brushing their front teeth - the incisors.

GEORGE. Would it be possible to use the brush and always keep the arrow pointing down?

HARRY. Certainly. If you knew you were going to get a wet arm it would be possible to drop your elbow a little and still get at those incisors.

IAN. I can't see anything against that arrangement for the working of the turbine.

GEORGE. Good. Now what about sealing. Do we need to keep water out of the mechanism or can we use it as a lubricant?

KEN. Hard water is going to foul it up in a couple of years.

HARRY. That's all right as long as the life can be increased by a simple modification.

IAN. Complete immersion will waste a lot of power.

GEORGE. All right, we'll have a wet area round the turbine and a humid area round the mechanism: water is the lubricant. Is there anything in the overall design that we haven't dealt with, apart from the connection to the tap? Good. Now, I think we must reject the idea of fitting the inlet pipe to an adaptor on the tap every time the brush is used: it's too inconvenient and expensive. We'll go for a semi-permanent fitting to the tap and people can get their cold water from the turbine outlet. The tap fitting must be made without tools of any kind; we'll use a large, easy-to-grip head on any screw that's used.

HARRY. Shouldn't the fitting be fairly easy to take off, so the tap can still have a hair washing spray fixed to it?

GEORGE. Yes, I think so but, if that's not possible, you might think about making it easy to fit something to the turbine outlet pipe. Now, how do you want to divide the work up, Harry?

HARRY. I suggest four basic areas: one, tap to turbine inlet; two, turbine nozzle, wheel and outlet; three, mechanism from turbine shaft to brush-head spindle; four, bearings and internal details of casing in consultation with Lawrence on industrial design.

GEORGE. That seems sensible but I don't really want to involve five different people. Ian, you've done a lot of work on the turbine and mechanism already: I'd like you to carry on with those and come to see me when you need some help with the detail design. There's a new chap called Mike Elrood who could do both the tap to turbine inlet and the casing, I think. Lawrence, you ought to be consulted about the connection to the tap as well, and if Mike does all the non-moving parts you only have him to deal with about details. Harry?

HARRY. That'll suit me very well. I'll brief Mike on what's happening. Are you definitely assigned to this project, Ken?

KEN. Yes, I'd like to be kept informed of everything that affects the production. If we get this order, it's very useful if I already know exactly what the thing will look like. Fred's already quite pleased that there are no electrical components.

GEORGE. So am I; it cuts down the communication problem considerably. There are still only four people and myself actively involved. Right, now to get the tender out on schedule I'd like to see some layout drawings by the end of next week please Ian. And Harry, will you tell Mike the same? We'll have a progress meeting about then.

CHAPTER 18

Solution Development

The sixth phase in the synthesis of a new product is that of making decisions on items of increasing detail, in consultation with the other people affected. During this phase many things are happening together and the exact nature of the work will depend very much upon the commercial arrangements with the customer. The work entailed in the overall design was divided up during Phase 5, and several quite distinct tasks have to be carried out before the preparation of the information required for the manufacture, assembly and testing of the product can begin.

The process described in Phase 4 is now carried out on a smaller area but to a greater depth. Generally, there is a pattern of events that is repeated as the work is progressively focussed on different parts of the overall design:

characteristic of experience is the habit of mentally evaluating several possibilities before stating a proposed solution.

An evaluation of the proposals for the bicycle drive could lead to the decision that gearing and shafting were too complicated, heavy and expensive and that a belt was not strong enough for the power and speed involved. The chain and sprockets must now be analysed so that details can be settled.

There is always some known data that help to determine certain important sizes, etc., but, almost invariably, there is no direct method of calculation. Design is *not* a process of substituting values into a formula and calculating the required unknown quantity. Real design work is nearly always iterative. In other words, when a certain solution is examined the analysis shows that it is not entirely satisfactory; the

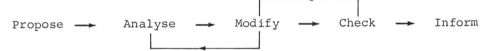

Propose $\rightarrow$ Analyse $\rightarrow$ Modify $\rightarrow$ Check $\rightarrow$ Inform

The person(s) concerned with a particular part of the product will first propose an outline of a solution to the particular detail problem on which they are engaged. This proposal may take the form of an idea in someone's mind, a verbal or written statement or a visual presentation, or some combination of these.

For example, whoever was first concerned with transmitting power from the foot-operated cranks of a bicycle to the back wheel might have proposed gearing, chain and sprockets, shafting, belt and pulleys, etc. Engineering students should be able to visualize how any of these methods could be employed, although a sketch would make the actual proposal clearer to the "man-in-the-street". Once proposals have been made, any evaluation is usually (unlike Phase 4) inclined to be quantitative. An experienced designer or small team would only put forward proposals worthy of serious consideration. Indeed, the

first solution is modified in a way that is expected to make it more satisfactory, and it is re-analysed. The process is repeated until the result is satisfactory.

Quantitative analysis is usually a process of making certain simplifying assumptions about the physical system and then using either an exact or a numerical or a graphical method to determine the required quantities. Alternatively, measurements may be made of the behaviour of full size or dynamically similar models. For example, a link in the bicycle chain may be analysed by making the simplifying assumptions that it is a perfectly uniform piece of metal that is not strained beyond the elastic limit and that the tensile force in the chain varies with a simple cyclic pattern. The link could then be checked to see that stresses and deformations were acceptable and fatigue tests on similar specimens could be used to predict the life

147

Halworths ATB project

Design Calculations

Designer: I. Montague

References:
(1) Kenpe's Engineers Year Book
(2) Company data sheets on synthesis of mechanisms

Turbine wheel

Specification: Speed(N) 1500 rev/min

Head(H) 2·5 m

Power(P) 2-6 W   say   3·5 W at max. efficiency

Specific speed $N_S = N\sqrt{P}/H^{5/4}$

$$= 1500\sqrt{·0035}/2.5^{5/4}$$

$$= \underline{28·2 \text{ rev/min}}$$

Reference (1) suggests Pelton wheel with one jet is suitable.

Assume total efficiency is 90%   Power = 90% × Head × flow

∴ Total flow $= \dfrac{3.5}{0.9 \times 2·5} = 1·56 \text{ N/s}$

$$= \dfrac{1·56}{9810} \text{ m}^3/\text{s} = 0·159 \times 10^{-3} \text{ m}^3/\text{s}$$

$$= \text{Area of jet} \times \text{velocity of jet}$$

Velocity of jet $= \sqrt{2 \times 9.81 H} = 7 \text{ m/s}$

If diameter of jet $= d$ then $\dfrac{\pi d^2}{4} = \dfrac{0·159 \times 10^{-3}}{7} \dfrac{\frac{m^3}{s}}{\frac{s}{m}}$

$$d^2 = 28·9 \times 10^{-6} \text{ m}^2$$

$$\underline{d = 5·38 \text{ mm}}$$

Let Pitch circle diameter of wheel $= D$

Reference (1) suggests ratio $\dfrac{D}{d} = \dfrac{210}{N_S}$

$$\therefore \underline{D = \dfrac{210}{28·2} \times 5·38 = 40 \text{ mm}}$$

As $\dfrac{D}{d} = 7.4$ Reference (1) suggests 18 buckets around the wheel and the buckets should have a radial height $b = 2.4 - 2.8 d$ say 13 - 15 mm and an axial width $a = 3.0 - 3.5 d$ say 16 - 19 mm.

FIG. 18.1

of such a link.

Once a suitable design has been established, the details must be communicated to the other people concerned; they may wish to comment on the proposals before they are finally adopted. One designer's proposals can affect other people in much the same way that the design factors (investigated in Chapter 15) affect the complete product. That is to say that there are some vital proposals that will completely constrain some aspect of another person's work, some important proposals that will still leave him some room to manoeuvre, and some proposals that may be taken into account if some advantage is thereby gained. It follows that someone must have overall control of the project and make decisions when the parties cannot agree amongst themselves.

There is a considerable overlap between this phase and phases 5 and 7. Some important details have already been decided when the overall design is produced and some quantitative analysis has been done in order to choose the most suitable general solution to the problem posed. In Phase 7 even the most minute details are settled and a considerable amount of quantitative analysis may be necessary, but it is during Phase 6 that the product is refined into a shape that will easily be recognized in the final manufactured product.

Once all the necessary decisions have been made, the product (or subassembly or component) is quite well defined but cannot yet be manufactured. However, it may be possible to estimate costs quite accurately. Some organizations are able to make a good estimate of costs long before this stage is reached but this is usually because they have made similar products in the recent past. The less familiar the organization is with the work that it is to undertake, the longer it takes to establish costs accurately. This problem is further complicated by technological innovation and, in recent years, by an increasing rate of inflation.

These problems should not be regarded as being outside the responsibilities of the engineer. Engineering is about making things TO SELL.

Phase 6 Making Decisions on Items of Increasing Detail, in Consultation with the Other People Affected

This exercise follows on from the Phase 5 exercises; the same organization has now reached Phase 6 of the automatic toothbrush project.

The Assistant Chief Designer has a meeting to discuss progress on the project. Study the calculations and sketches and note the decisions made at the meeting.

Prepare a layout drawing of the complete driving unit, i.e. a drawing which indicates every part of the assembly, to scale as far as possible, but with the less important details left to be settled at the assembly and detail drawing stage.

Compare your solution with those shown in Chapter 19 (*Figure 19.1*).

Scene: The office of an Assistant Chief Designer

Present: George   - Assistant Chief Designer
Harry   - Design Engineer
Ian   - Designer (Mechanical)
Ken   - Production Engineer
Lawrence - Industrial Designer
Mike   - Junior Designer (Mechanical)

This meeting was arranged since the Phase 5 meeting.

GEORGE. I thought it would be useful, now that we have a basic layout with some definite ideas of sizes et cetera, to review the work that's been done and plan the remaining work. Where's the best place to start, Harry?

HARRY. The turbine, definitely: it affects everything else. Now neither Ian nor I could find anything useful in the library on impulse turbines of this size. There's plenty on propeller type reaction turbines as used in flow meters but their main design criterion is to use as little of the available head as possible; so that was no help. We decided to use the analysis that works for full size machines. We'll almost cert-ainly have a much lower efficiency because of manufacturing inaccuracies and our proportionally higher friction losses; on the other hand our nozzle should be more efficient and give a more compact jet. I think Ian can best explain the details.

IAN. Here are my calculations. You can see on sheet 1 (*Figure 18.1*) that, to get the necessary flow of water, our jet diameter will have to be five point three eight millimetres. The pitch circle diameter of the wheel, for the correct power at the design speed, should be forty millimetres.

Ideal shape of bucket (Reference (1))

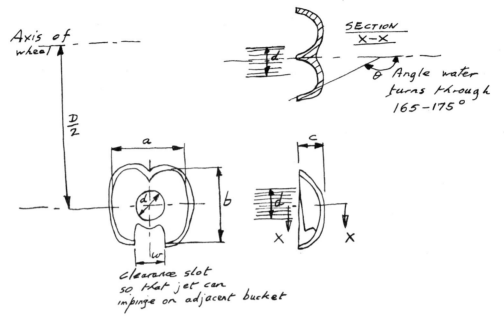

Slot width   w = 1.2d   say 6.4 mm
Bucket depth c = 0.8 - 0.95d   say 4.5 - 5 mm

Arrangement of buckets

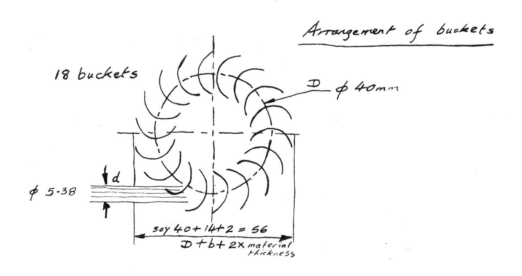

FIG. 18.2

LAWRENCE. Yes, you mentioned that the other day, but where did your figure of seventy something come from?

IAN. We'll get there is easy stages. On sheet 2 (*Figure 18.2*) I've sketched out the recommended shape of the little buckets that the water hits. They should be about sixteen millimetres in the radial direction so the outer diameter of the wheel comes out at fifty-six millimetres.

LAWRENCE. Is there no way of reducing that?

HARRY. The short answer is "yes" but that won't improve the performance or the cost. The point is that the average speed of a bucket ought to be half the speed of the jet in order to get the best efficiency. We could have two jets and reduce the wheel diameter but that would mean an increase in the speed of rotation.

KEN. A smaller wheel would be more difficult to manufacture as well.

LAWRENCE. Ah well, it'll be a challenge for me I suppose.

IAN. You can see the basic arrangement of the buckets at the bottom of sheet 2. Eighteen was the number that would be used on a full size turbine but we thought we would cut that down to sixteen to reduce the extent to which they mask each other. Now, you can see the geometry of the nozzle at the top of sheet 3 (*Figure 18.3*) and I've incorporated that in a casing for the turbine at the bottom. That's where the seventy four millimetres comes from, Lawrence.

LAWRENCE. Yes, I see why you need so much radial clearance at the top and bottom, but do you need it at the sides? The toothbrush is never held so that the water could collect there.

HARRY. That's true. I think we could give you some space to use there.

IAN. Right, I'll make a note of that.

MIKE. One other thing: does the internal diameter of the water supply tube have to be twenty millimetres all the way along?

HARRY. Ideally, yes Mike, but the tap itself isn't likely to be that size so I think we only need twenty millimetres near the nozzle.

GEORGE. You'll have to try a lot of these things out on a prototype, if we get the order. What's next?

IAN. Sheet 4 (*Figure 18.4*) is just my thinking on paper really about the kinematics of the mechanism. At the top of sheet 5 (*Figure 18.5*) are four positions of something like the final proposal, and you can see how I would fit it in at the bottom. At the top of sheet 6 (*Figure 18.6*) is the proposed mechanism, full size. I just drew a circular casing round it but only a quarter of that area is required.

KEN. What are those spring clips: are they metal?

IAN. No, I've just drawn them like circlips for quickness. I'm no expert on plastics but presumably there is some suitable fastener on the market.

KEN. It's not so easy but we could sort it out with your detail designers.

IAN. Yes, as long as you're happy about the relative positions of the links for assembly?

KEN. That seems all right.

HARRY. Good. Now Mike had a look at the moving parts and made a few suggestions about supporting them. Is that the lower drawing on sheet 6?

MIKE. Yes, Ian and I thought we'd keep all this part of the design together. I must apologize for the casing: it's just a first rough drawing that was all I had time to do after finishing the tap connection. You see that we need another join.

GEORGE. That's all right Mike; you've got time to knock it into a better shape.

LAWRENCE. As you don't really need so much clearance on either side of the turbine wheel, you could tidy that part up considerably. Presumably, the water outlet tube serves as the handle?

MIKE. That's the general idea; again we need a model to establish the best internal geometry but I shouldn't think it will constrain you very much.

LAWRENCE. The centre line of the brush-head follows through to the water outlet I see. Is that essential?

MIKE. Oh no, it's an ergonomic problem for you; I had to show something and I thought that the outlet pipe centre should be on that side for easy handling. For easy water outflow it would be better nearer the wheel axis.

LAWRENCE. I'm sure we can find some compromise solution.

Halworth's ATB project

Ideal shape of nozzle (Reference (1))

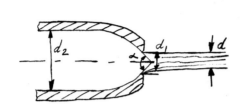

$d_1 = 1.2 - 1.4\ d$ say 7mm
$d_2 = 3 - 4\ d$ say 20mm
$\alpha = 60 - 90°$ try 90°

Suggested arrangement of complete turbine with 16 buckets to reduce interference

SCALE 1/1

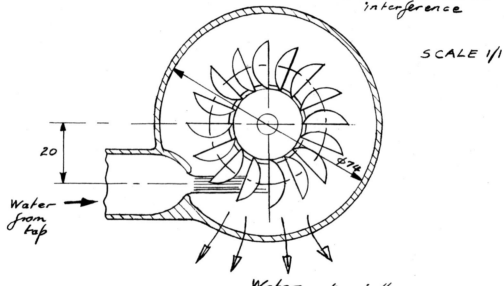

20

Water from tap

Water out axially

Comments: Very large diameter needed to avoid flooding of turbine — casing cannot be gripped at this cross-section.

FIG. 18.3

KEN. Sorry if you've already though of this: the turbine shaft should have a shoulder on it in order to ensure that the wheel clears the end of the bearing.

MIKE. Thanks. As I said, it was a very quick drawing.

HARRY. Good. Now can we see your sketches of the tap connection, Mike?

MIKE. Yes. I've labelled my sheets with Roman numerals to distinguish them from Ian's. Sheet 1 (*Figure 18.7*) just shows my original ideas on how to fit all sorts of taps and how to hold the fitting in place. I've refined them on sheet 2 (*Figure 18.8*) and then sheet 3 (*Figure 18.9*) shows a layout drawing of the complete connection.

GEORGE. I see you've mentioned avoiding a possible patent for the clip - we'll have to be careful that there are no patents on a connection like this.

KEN. I've got one on the end of my garden hose that certainly resembles this but I wouldn't have thought that there's much you could patent.

HARRY. I'll have a chat to our patents people, just in case.

GEORGE. Fine. Let's see what remains. Ian and Mike are going to tidy up the layout on Ian's sheet 6 (Fig.18.6) and

Lawrence can do a few sketches now that the basic layout is settled. Assuming that the tap connection is all right, I think we can get that costed as it stands. I know what I wanted to ask you Lawrence: are you going to suggest any awkward colours or textures? You know the sort of things that put the cost up.

LAWRENCE. I thought about using transparent material around the turbine. It would appeal to children to see it work working.

KEN. Brilliant idea - but it'll take the cost up a bit I think.

HARRY. It's the first proposal that has though, and it could just clinch the order for us.

GEORGE. I agree. Let me have a decent layout drawing and the industrial design sektches for the costing boys as soon as you can. Then Harry can write the technical part of the tender and we'll hand it over to the Director of New Projects. He's managed to persuade his Masters, as he calls them, to let him quote on the assumption that we produce more than just the initial batch. So, with a toothbrush that uses no batteries, has a fascinating turbine wheel whizzing round in full view, and comes at a bargain price, I'll be very surprised if we don't get the order.

Halworth's ATB project

Oscillating Motion Mechanism

Specification : Input speed (N) 1500 rev/min
Power, say 3·5W
Output – tip of bristle 10mm long
is to move through 8mm

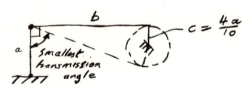

$\sin \theta = \frac{4}{10}$ ∴ $\theta = 23°\,35'$

Let brush head be driven by crank of radius a.
Reference (2) states that linear movement of crank must
be twice radius of driving crank – c

∴ $c = a \times \frac{4}{10}$

For good transmission of force make connecting crank
near 90° to driven crank when the latter is at mid-stroke

Reference (2) suggests that transmission angle should be
between 70° and 110°.

Hence $\tan 2\theta = \frac{4a}{10} / b$
or $b = \frac{4a}{10} \tan 70$
$= 0·9988\,a$

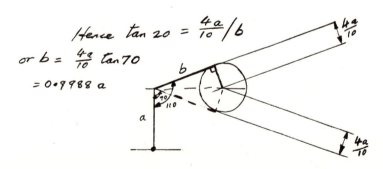

FIG. 18.4

Halworth's ATB project

Even better transmission angles could be obtained
by making $b > 0.9988a$ say $b = 1.5a$
$c = 0.4a$
then locus is:

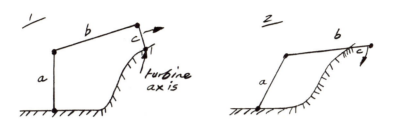

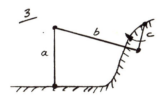

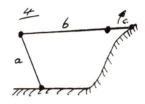

Worst transmission angle occurs in **2** now.
This can be improved by increasing b or altering
the distance between the fixed pivot points.
b cannot be so large that the driven crank has no
clearance space inside the casing.

Let distance from turbine axis to brush head
axis = e and let radius of casing in this area = r

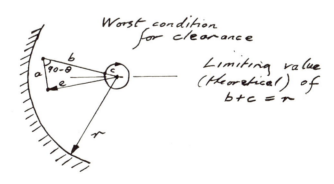

Worst condition
for clearance

Limiting value
(theoretical) of
$b + c = r$

FIG. 18.5

Halworths ATB project

Suggested arrangement of mechanism

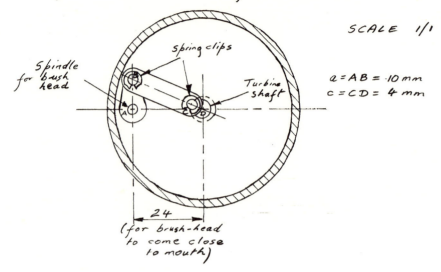

SCALE 1/1

Spring clips

Spindle
for brush
head

B

A

Turbine
shaft

D

$a = AB = \cdot 10\,mm$
$c = CD = 4\,mm$

24

(for brush-head
to come close
to mouth)

Plan view of moving parts, bearings and water
passages

SCALE 1/1

WATER
IN

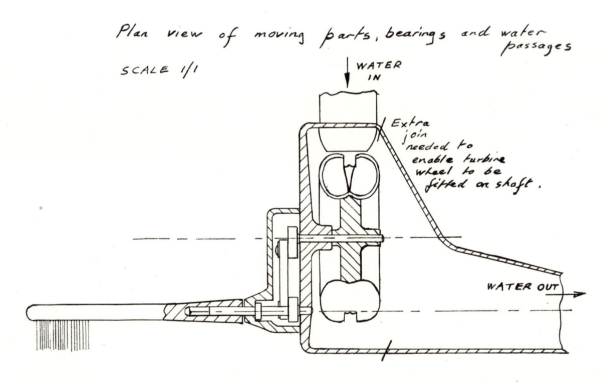

Extra
join
needed to
enable turbine
wheel to be
fitted on shaft.

WATER OUT

FIG. 18.6

HALWORTHS ATB PROJECT

DESIGNER : M. ELROOD

CONNECTION TO TAP

SPECIFICATION : EASY FITTING TO
ALL TYPES OF TAP.
TO TOOLS TO BE USED.
PROBABLE BATHROOM/KITCHEN TAPS :
1. 12 mm ($\frac{1}{2}$ inch) ROUND END
2. 12 mm ($\frac{1}{2}$ inch) OVAL END
3. 18 mm ($\frac{3}{4}$ inch) OVAL END
4. 12 mm ($\frac{1}{2}$ inch) NON-CYLINDRICAL

INTERNAL SHAPE NEEDS TO DEFORM TO ACCOMODATE
DIFFERENT SIZES :

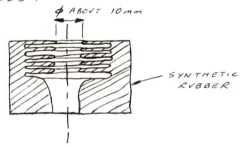

ϕ ABOUT 10mm

SYNTHETIC
RUBBER

GRIP CAN BE INCREASED BY MEANS OF A BAND
TIGHTENED AROUND THE RUBBER.
"JUBILEE" TYPE CLIP IS PATENTED?
NOT EASY TO INCORPORATE HAND OPERATED SCREW.

PIPE VICE SORT OF CLAMP COULD BE USED !

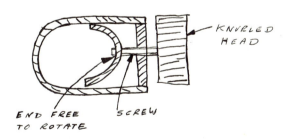

KNURLED
HEAD

END FREE
TO ROTATE

SCREW

FIG. 18.7

TAP FITTING NEEDS NOZZLE FOR TUBE AND
LEAD IN FOR TAP. ALSO RECESS FOR CLIP.

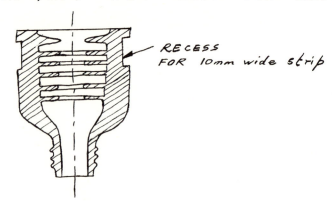

RECESS
FOR 10mm wide strip

CLAMPING STRIP SHOULD BE FLEXIBLE TO ALLOW
FOR OVAL TAPS. STATIONARY THREADED PART COULD
GO INSIDE THIN STRIP AND MOVING PART COULD
SLIDE OVER IT.

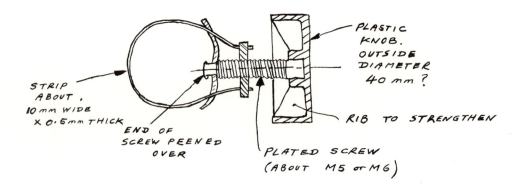

STRIP ABOUT.
10mm WIDE
x 0.5mm THICK

END OF
SCREW PEENED
OVER

PLASTIC
KNOB.
OUTSIDE
DIAMETER
40 mm ?

RIB TO STRENGTHEN

PLATED SCREW
(ABOUT M5 or M6)

PEENED END SHOULD BE FAIRLY SMOOTH SO
THAT RUBBER IS NOT PIERCED.
TUBE FROM CONNECTION TO TURBINE NOZZLE
NEEDS TO BE CEMENTED AT EACH END.
TUBE MUST BE FAIRLY THICK-WALLED SO THAT
FLOW IS NOT EASILY RESTRICTED.
KNOB COULD HAVE SIMPLE INSTRUCTIONS MOULDED
ON HEAD.

FIG. 18.8

HALWORTHS ATB PROJECT

TAP FITTING

THIRD ANGLE PROJECTION   SCALE 1/1

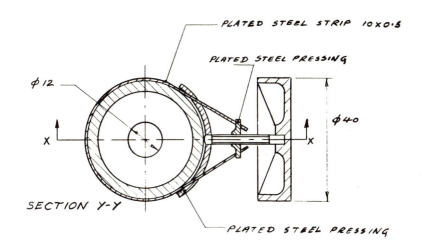

PLATED STEEL STRIP 10×0.5

PLATED STEEL PRESSING

φ12

φ40

X — X

SECTION Y-Y

PLATED STEEL PRESSING

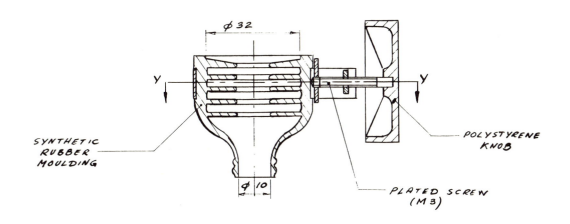

φ32

Y — Y

SYNTHETIC RUBBER MOULDING

φ10

POLYSTYRENE KNOB

PLATED SCREW (M3)

SECTION X-X

M. Elwood

FIG. 18.9

CHAPTER 19

Solution Execution

The seventh phase of a design project is that of directing the manufacture of the required quantity. It was stated (in Chapter 10) that the manufacture of a component or the correct assembly of a set of components is facilitated by engineering drawings. The work involved in translating the layout drawings into working drawings is considerable in magnitude but often somewhat routine in nature. Quite often there are calculations still to be done but they are likely to be rather less interesting than the earlier analysis that established the major quantities involved in the problem.

For these reasons there is a tendency for the work to be handed over to detail designers and draughtsmen who have not previously been involved in the project and can only regard it as another job, similar to others that they have seen in the past. The worst that can happen is that the more senior members of the design team lose interest in some aspects of the problem at this stage and, for various reasons, some apparently minor decision is taken by a junior employee which adversely affects the whole product.

Of course, there must be a sharing of responsibility at this phase; the work would take too long if other people were not brought in. It is the supervision of this work that is sometimes at fault. A good overall design can be ruined by poor detail design. The most successful organizations are well aware of this and take steps to ensure that, at least, all detail work is properly checked, at best that the new personnel are well-motivated by familiarizing them with all those parts of the project that could be of interest.

There are different ways of organizing the assembly and detail drawing work. Some drawing offices prepare a very accurate assembly drawing which is then translated into the required detail drawings of the various components. Other organizations refine the layout drawings into something approaching the final assembly drawing, produce the detail drawings from this, and then make a completely accurate assembly drawing from the detail drawings. Draughtsman's time is valuable and the first method may produce a saving as long as the assembly drawing does not have to be drastically modified before issue to the manufacturing departments.

At this stage of the project the production engineers and development engineers may become deeply involved and may suggest changes once prototypes have been built and tested. Simple products are usually modified by issuing a revised drawing. More complex products often have many completely different drawings for the prototype and production models.

The detail designer must have a considerable knowledge of manufacturing processes and is sometimes made responsible for planning the manufacturing operations. In other words, he decides the form of the raw material and exactly what is to be done to it in order to produce the component shown on his drawing. Alternatively, the designer/ draughtsman concentrates on drawing something that can be made but the actual production details are left to a specialist who is attached to the manufacturing departments.

Manufacturing processes are an important constraint on the designer/ draughtsman and components are often drawn out in a particular way mainly because they are to be made by a certain process. Fundamentally, there are only three ways of shaping material:

1. Start with too much and cut away that which is not required,
2. Start with the right amount and deform the material to give the required shape,
3. Fill a space of the required shape with a liquid material which will later solidify.

The variations and combinations of these methods are legion but, generally speaking, they are in order of suitability for small quantities of complex shapes. The detail design is greatly affected by the quantity to be manufactured.

160

Finally, the detail designer is involved in the choice of materials to be used. Whereas manufacturing processes have developed considerably in a decade, new materials have been introduced at a bewildering rate. Engineers deal with two basic types of material: those with small compact molecules, such as metal and liquids, and those with long molecules (high polymers) such as rubbers and plastics.

The number of applications for the latter type has increased rapidly and a new outlook is spreading through industry: increasingly the question is asked "Is there a non-metallic material that it would be an advantage to use?". This certainly does not mean that metals are obsolescent; there are many applications where a metal is the only possible material and this will be true for the foreseeable future.

Phase 7 Directing the Manufacture of the Required Quantity

These exercises follow on from the Phase 6 exercises; the same organization has how reached Phase 7 of the automatic toothbrush project.

1. Referring to Chapter 18 only, make sketches of the external appearance of the proposed automatic toothbrush.
Compare your sketches with Figs. 19.2, 19.3 and 19.4.

2. Suggest ways in which the detail design work involved in the layout drawings Fig.18.9 and Fig.19.1 could be organized.
Compare your suggestions with the decisions taken at the meeting arranged by the Assistant Chief Designer - Scene 1.

3. Scene 2 is a short discussion on the performance of part of the prototype toothbrush. By picturing yourself using the complete product, list ways in which poor detail design could affect its operation.
Compare your ideas with those expressed by Andrew in Scene 3.

4. Make subassembly and detail drawings of the driving spindle and its 10 mm radius link. Incorporate a modified end cross-section on which the brushhead will fit so that the bristles are always pointing towards the user.
Compare your drawings with Figs. 19.6 and 19.7.

5. The layout drawing in Fig.19.1 shows a system of dividing the casing which is, deliberately, far from ideal. Suggest a more suitable system to improve appearance, function and assembly.
Compare your system with that proposed by Mike in Scene 3.

Suggestions for further exercises
Students who wish to proceed further with the detail design of the automatic toothbrush will find sufficient information in Chapters 18 and 19 to tackle many of the parts in the turbine wheel and mechanism subassembly.
The author has no doubts that this particular design could be made to work but much of the material relating to this project has been introduced to illustrate the process of synthesizing a new product; there is no suggestion that a company would necessarily adopt this solution.

Scene 1: The office of an Assistant Chief Designer

Present:
George   - Assistant Chief Designer
Harry   - Design Engineer
Ian   - Designer (Mechanical)
Ken   - Production Engineer
Lawrence - Industrial Designer
Mike   - Junior Designer (Mechanical)

GEORG . As I think most of you know, we won the order for an automatic toothbrush for Halworths and we are just about to initial a contract with them for the supply of a test marketing batch of twenty thousand. The contract is going to stipulate that delivery is to commence on the first of September so that gives us just six months. Now I hope you've all had a chance to refresh your memories on the notes and drawings that you produced for the tender. If you could find your final layout drawing, Mike we can see what's to be done.

MIKE. This one, ATB 5? (*Figure 19.1*)

GEORGE. That's it. Now I've jotted down a few of the things we have to deal with. The first one is making a prototype; Halworths will want to comment on it before the production model is finished. Will you give that top priority please, Harry?

HARRY. Right. Are most of your ideas embodied in ATB 5, Lawrence?

LAWRENCE. Quite a lot, but I'd still like to try an exit orifice that's

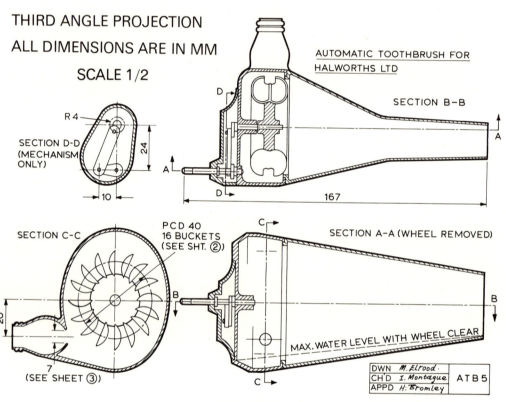

THIRD ANGLE PROJECTION

ALL DIMENSIONS ARE IN MM

SCALE 1/2

R4

SECTION D-D
(MECHANISM
ONLY)

24

10

AUTOMATIC TOOTHBRUSH FOR
HALWORTHS LTD

SECTION B-B

D

D

A

A

167

SECTION C-C

PCD 40
16 BUCKETS
(SEE SHT. ②)

C

SECTION A-A (WHEEL REMOVED)

B

B

20

MAX. WATER LEVEL WITH WHEEL CLEAR

7
(SEE SHEET ③)

C

DWN	M. Elrood	
CH'D	I. Montague	ATB 5
APPD	H. Bromley	

FIG. 19.1

directed slightly away from the user. And I'm not at all keen on that chimney-pot shaped nozzle in the top view.

HARRY. I'm inclined to agree about both. Where are your designs?

LAWRENCE. There are four here. I use my initials and a departmental numbering system to identify them. LB 0338 (*Figure 19.2*) and LB 0339 (*Figure 19.3*) are a couple of early ideas. LB 0340 (*Figure 19.4*) is a first accurate drawing, and LB 0341 (*Figure 19.5*) is the polished presentation that Halworths saw. The only other difference between that and ATB 5 is the transparent inserts near the turbine.

HARRY. I think the model shop can work directly from your drawings but we shall need some detail drawings of the turbine wheel and mechanism for the tool room. Who can I have to help, George?

GEORGE. Ian and Mike for the time being, and there's Norman Cotton who's nearly finished another job: he's quite an expert on plastics in mechanisms. Don't produce a tap connection for the prototype, use whatever you can that's already available.

KEN. What's happening about that? Are you thinking of subcontracting it?

GEORGE. Thinking, yes. I'd rather make it ourselves but the Works Manager is concerned about the effect that could have on another order that uses synthetic rubber mouldings. We'll have to wait and see. However, I think you could go ahead and produce an assembly drawing, Harry, so that it's in stock, so to speak. Now the next priority is production planning. Ken?

KEN. ATB 5 seems to be mostly plastic mouldings with odd pieces of metal for

the shafts. I'll arrange a meeting with the appropriate people and see if they spot any snags with this layout but personally I think it's just a question of phasing the work to fit in with the other jobs. The most complicated part is the turbine wheel; is there a chance that we could get to work on that quite soon? It's a well-tested design; you shouldn't have to alter it when you test the prototype.

IAN. I'm not so sure. We've no information on how well one of this size will work and we may need to increase the moment of inertia with a metal insert for the plastic production model. It's almost impossible to predict the loads so we shall have to get a few people to clean their teeth and see how badly the wheel slows at peak loads.

GEORGE. Can you give that absolute priority then, Harry. Get the tool room on it this week, if you can, and we'll find out if the turbine works so Ken knows what he's going to deal with.

KEN. Thank you. Just one other thing: who's going to make the inlet tube?

GEORGE. That's easy: you buy that in reels and chop it up to the length required. Lawrence has the details. O.K.? Now I'll call another meeting as soon as there's something to report and, in the meantime, I'll be issuing a schedule of events and target dates so we can start to think about how much more manpower we shall need.

HARRY. I'll come and see you later, Lawrence and talk about materials and colours for the casing. Ian, will you drop everything and draw up a turbine that the tool room can produce and a suitable nozzle. Mike, give Ian a hand with the drawings when he's ready, but you may as well start that fitting assembly while you're waiting.

Scene 2: A Testing Area

Present:
George - Assistant Chief Designer
Harry - Design Engineer
Ian - Designer (Mechanical)
Ken - Production Engineer
Norman - Detail Designer

HARRY. Well, there you are. The turbine works all right but it's very inefficient. We seem to be getting just over two watts out of a theoretical six watts that's going in. When the turbine gets flooded or there is a big peak load it can slow down to rest in just over a second. I'd say that the moment of inertia is about right.

GEORGE. Looks like it. What's it made of?

IAN. That's epoxy resin. There's a greater mass there than we'd probably have in the production models.

NORMAN. It's not a very accurately made wheel. I think we'll get more power out of a properly moulded one.

KEN. Yes. So it looks as if there will have to be a metal insert to get the same moment of inertia?

163

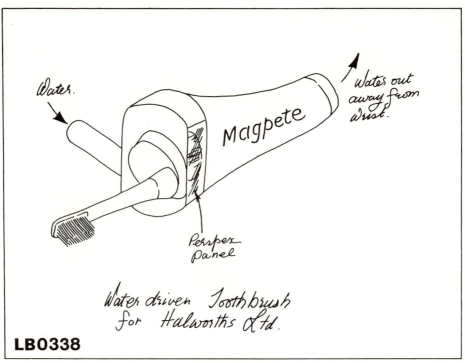

LB0338

FIG. 19.2

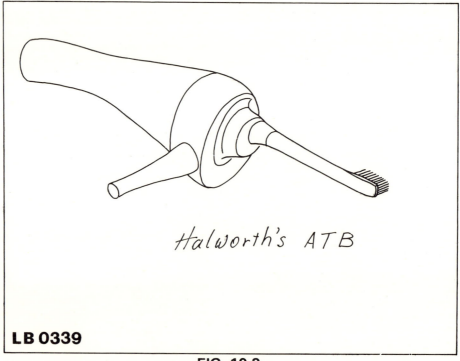

LB 0339

FIG. 19.3

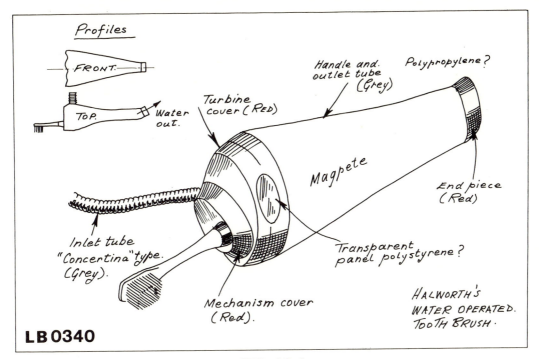

FIG. 19.4

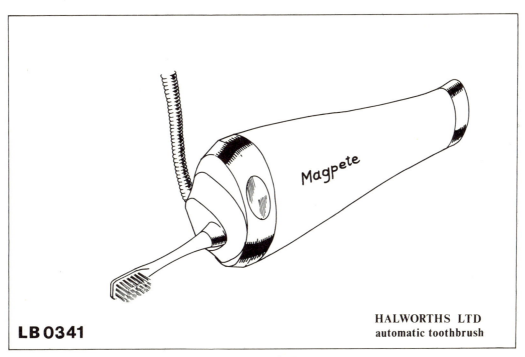

FIG. 19.5

HARRY. We won't if we can possibly
avoid it and keep the costs down. Do
you think you could redesign it Norman,
so that more of the mass is away from
the centre?

NORMAN. I'll try. The other thing
that's slowing it down is the bearing
and journal materials. If we could use
p.t.f.e. or something like it, it would
be a lot better.

HARRY. See what you and Ian can devise.
We don't particularly want another part
to insert.

GEORGE. Is the nozzle satisfactory for
the pressure you're using?

HARRY. Seems to be. It gives a nice
clean jet and we should be able to get
the geometry just as accurate in polystyrene and a more efficient turbine
wheel in polyethylene; we may be able
to cut the flow down a bit. It strikes
me as rather a lot to dispose of.

KEN. How soon could we see the detail
drawing of the wheel?

NORMAN. That's what I'm concentrating
on at the moment.

GEORGE. Let's say next week then, Ken.
How's the rest of the prototype, Harry?

HARRY. Tap connection and tube: no
problems. Same with the mechanism.
What's happened to the casing, Ian?

IAN. The model people are also using
epoxy resin and wooden formers to get
the general shape. We hope to fit this
turbine wheel into it and bond on different nozzles to get the best compromise.

GEORGE. Good. Now, as soon as the
prototype is working satisfactorily
we'll be able to get a few more draughtsmen working on the details. Then I
think it would be a good idea to review
progress once Halworths have seen the
prototype, and get the Director of New
Projects along to help to put everyone,
especially the new people, in the
picture.

Scene 3: The company's conference room

Present:
Andrew	-	Director of new projects
George	-	Assistant Chief Designer
Harry	-	Design Engineer
Ian	-	Designer (Mechanical)
Ken	-	Production Engineer
Lawrence	-	Industrial Designer
Mike	-	Junior Designer (Mechanical)
Norman	-	Detail Designer
Paul	-	Tool Designer/Draughtsman
Richard	-	Draughtsman
Simon	-	Drawing Office Apprentice

HARRY. It's very pleasant to see all
the members of my present team gathered
together in one place; I'd normally
have a long walk to see so many of them.

GEORGE. Yes. Well, I'm glad we could
find a convenient time for the Director
of New Projects to give you some background information, Andrew?

ANDREW. Our design won this order
mainly, I think, because it's different.
Other companies showed Halworths battery-
driven brushes which were quite good
and easy for children to use but they
were looking for something a bit more
than that. Our brush is, if we're hon-
est, not so convenient to use but it's
got tremendous potential child-appeal.
Kids are used to playing with battery-
driven objects these days so a little
water turbine has a great advantage.
The initial order is for twenty thousand
but I shall be very disappointed if we
don't get future orders of at least
fifty thousand a year. Whether we do
or not depends to a great extent on all
of you. Now Halworths have seen the

prototype working and they only had one
or two minor complaints; George will
deal with those later I expect. However, let me give you my private nightmare of the things that could go wrong
at this stage. Just picture a little
boy or girl who's just been given one
of our toothbrushes. Father fixes it
on the tap for the child, turns it on
fairly gently, according to the printed
instructions and splosh! the fitting
comes shooting off producing dampness
and depression all round. He tries
again with the same results and the
child learns a few new words. Eventually, being a bit of a handyman, he
finds a suitable clip and succeeds in
making the tube stay on. He's already
toying with the idea of punching the
manager of the local Halworths on the
nose; fortunately for me he doesn't
know how much I had to do with it - yet.
The toothbrush starts working, we hope,
but stops every time that the child
applies more pressure. Having brushed
his left side molars using the right
hand, he changes sides and promptly gets
water right up the sleeve of his pyjama

jacket. Eventually, after much stalling, the child's teeth are done and Father says: "Let me have a go". He tugs at the brush-head which won't come off so he tugs again and its stem breaks. Downstairs to the garage to put the broken end in a vice and Father is feeling displeased with us once more. He succeeds in removing one now useless brush-head and repairs to the bathroom where he gingerly pushes a new head on to the spindle. This one is loose because it's in a different position relative to the spindle. Father adjusts it, turns on the tap and starts brushing. He doesn't think the action is sufficiently vigourous so he turns the tap on more and pushes the brush-head into the furthest corner of his mouth. Something fails inside the mechanism and Father is drenched into the bargain. Now by this time, he's not really interested in having his money back; his one ambition is to see my head on a plate, and of course he won't be the only father affected. I might be able to cope with one homicidal maniac storming into my office but I really don't want more than a dozen: that would be too much. So please, when you're doing the detail work on this project, think of me and my nightmare: it need not come true.

GEORGE. Of course, the other extreme is that the thing is so expensive to produce that we don't make a profit on the deal.

ANDREW. Don't! That's too horrible to contemplate.

GEORGE. Do you want to run through the comments that Halworths made on the prototype?

ANDREW. No, you do that. I shall have to be getting back. Is there anything any of you want to ask me?

PAUL. This problem with water going over people: don't you think they'll get used to using the brush in such a way that it doesn't happen?

ANDREW. Oh yes, I'm sure they will but what we must do is make sure that it's easy to do that; even if it means an extra part and a few extra pence.

GEORGE. Anything else? Well, thank you very much Andrew. Now, can we have a progress report, Harry? I'll deal with Halworth's comments as they arise.

HARRY. Right. If you look at the layout drawing, ATB 5 and Lawrence's rough sketch, LB 0340, I'll run through what's happened so far. Norman's finished the detail of the turbine wheel and Ken's people have looked at the first issue.

KEN. There are a few details to clear up with Norman and then the tooling to be considered. We could make it in a high density polyethylene and get about the right mass.

HARRY. Mike made a preliminary assembly drawing of the tap fitting and, now that we're manufacturing it ourselves, Lawrence has improved its appearance, and Mike and Simon are doing the production drawings.

GEORGE. Let's have a few prototypes as soon as you can. You saw how concerned Andrew was about drenching people.

HARRY. Yes, that's in hand. We've got some of that concertina tubing that Lawrence suggested and it bonds on to synthetic rubber very well. The mechanism: Norman's working on that now that he's finished the turbine for the time being.

NORMAN. I've a couple of drawings here, actually. SA 7918 (*Figure 19.6*) is a subassembly of the driving spindle and its link. D 15770A (*Figure 19.7*) is a detail of the spindle.

KEN. That's quite useful; will they be issued soon?

IAN. They're still to be checked and approved but I'll send you copies when they have been.

GEORGE. Halworths are going to buy the brush-heads from the Porcupine brush company. They'll need copies of those drawings.

HARRY. That's mainly why I got Norman working on them at this stage. Lawrence also needs to contact them about the general shape and colours.

LAWRENCE. I've been in touch already. They use high impact polystyrene to set the nylon bristles in; we could use a similar material for the red parts of my sketch and keep the same surface texture.

GEORGE. That sounds sensible. Now, Halworths pointed out that, with our square-ended spindle, a child could put the brush-head on so that it was wrongly aligned with the outlet pipe. Also with a square rigid end the mating square hole tends to enlarge and eventually the brush-head won't stay on. I see you've changed the cross-section that fits into the brush-head so it can only be aligned one way. It'll still give problems with loose brush-heads.

NORMAN. The difficulty is that, with a new head and some sort of wedging

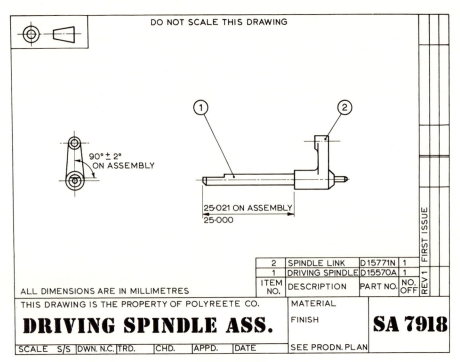

FIG. 19.6

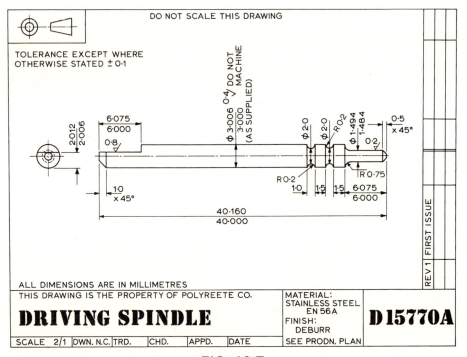

FIG. 19.7

action, it would be all too easy to exceed the ten newtons maximum force that's in the specification.

GEORGE. I see. Well, leave it like that for the initial batch; we may have to change it when Halworths have some customer reactions.

HARRY. O.K. Now I've divided the rest of the work in two. First, those parts which contain bearings, i.e. what Lawrence refers to in LB 0340 as the "mechanism cover" and "turbine cover". Second, the parts without bearings, i.e. the "outlet tube" and "end piece".

GEORGE. Halworths felt quite strongly about that. The prototype is very difficult to use without getting wet. We suggested a modified outlet tube but what they wanted, er: colour contrast. Perhaps Lawrence can explain . . .

LAWRENCE. The change in colour at the end gives a visual emphasis to the fact that something special happens there. I'm doing a coloured drawing at the moment, mainly for Halworths' benefit, but I don't think there's much doubt that they'll insist on a different colour at the water outlet.

HARRY. Ah well, it's convenient in many ways, I suppose. Now Ian and Paul are working on the mechanism and turbine covers. How far have you got?

IAN. We want to change the way that they are divided. The division in ATB 5 was fairly convenient for making the prototype but there's a better way. We'll have all the external surfaces and the left-hand bearing in one moulding; Lawrence is improving their appearance. The two right-hand bearings are

in a separate internal wall and the complete wheel and mechanism can be assembled with that. Then the subassembly can be inserted into the combined mechanism and turbine cover.

KEN. Good. That's a much better arrangement for production.

PAUL. In fairness to Mike, he thought of it first but rejected it for the prototype.

GEORGE. Now that will deal with another of Halworths' complaints. They thought the covers looked too fussy.

LAWRENCE. I've smartened them up considerably in my colour drawing.

HARRY. That just leaves the outlet tube and end piece. Richard's dealing with that . . .

RICHARD. It seems quite straightforward except for the material for the outlet tube. Lawrence has suggested Polypropylene. It's still a bit pricey.

GEORGE. Halworths thought the prototype tube was unpleasant to handle so I'd like to see an improvement there.

RICHARD. I'll sort it out with Harry when we've got a better idea of costs.

HARRY. Well, that's about all I think. It's coming along quite well and there are only a few minor snags.

GEORGE. Good, let's hope it stays that way! If Halworths sell as many of these as they think they will, our work will soon be appearing in half the bathrooms in the country.